SUCCESSFUL AIR CONDITIONING & REFRIGERATION REPAIR

BY ROGER A. FISCHER

TAB BOOKS Inc.

BLUE RIDGE SUMMIT, PA. 17214

FIRST EDITION

FIFTH PRINTING

Printed in the United States of America

Reproduction or publication of the content in any manner, without express permission of the publisher, is prohibited. No liability is assumed with respect to the use of the information herein.

Copyright © 1982 by TAB BOOKS Inc.

Library of Congress Cataloging in Publication Data

Fischer, Roger A.
    Successful air conditioning & refrigeration repair.

    Includes index.
    1. Refrigeration and refrigerating machinery—Main-
tenance and repair. 2. Air conditioning—Equipment and
supplies—Maintenance and repair. I. Title.
TP492.7.F57   621.5′6   81-9103
ISBN 0-8306-0015-9   AACR2
ISBN 0-8306-1281-5 (pbk.)

Contents

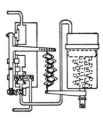

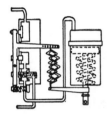# Preface

This is a practical book. It was written by a man who presently works in the field and has instructed in the field as well as in the classroom. This book is a reference manual that can be carried in your tool box and can be consulted when you have problems. Most of the information in this book was gained through work experiences and does not appear in any other text or reference manual.

When I was instructing, my students would bring many different books out on our class field problems. These texts and reference books were many pages with a lot of words that could not be related to the field. The books were theory with classroom lectures. The students asked me for a "hands-on" reference guide that can be easily carried. My hope is to fulfill those requests. Reader inquiries will be gratefully acknowledged and appreciated.

Chapter 1
Basic Electricity for Refrigeration

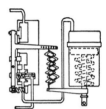

I want to cover in a practical way the failures and troubles that occur in the air conditioning field. If you were to analyze air conditioning failures, you would find that 80 percent of them are electrical failures. The remaining 20 percent of the problems are in the closed refrigerant system. With this in mind, I have focused 80 percent of this book on electrical malfunctions and 20 percent of this book on closed system failures. If you were to look at the books on the library shelves concerning air conditioning and refrigeration, much of the texts are not relevant to the field. They might have one or two chapters at the most on electrical failures. I have written this book so that any student or professional will be able to go out in the field with a few tools and diagnose problems.

With this in mind, I will start off with basic electricity. High-voltage transmission wires are carried on insulators on the top cross members of telephone poles or power poles. Voltage is quite high. It can be 4160 volts, 3600 volts, 2400 volts or higher voltages. The purpose of using such high voltage is to cut down on wire resistance line losses.

Figure 1-1 shows a step-down transformer with the primary attached to the high-voltage top cross bar conductors. The secondary of this transformer, which has a center tap neutral and two line wires, line 1 and line 2, is connected to the lower conductors on the bottom cross bars of the power poles. In the schematic, the secondary wires are leaving from the transformer and are wired to an electrical watt-hour meter that records power consumed.

From the bottom of the meter, the conductors are wired to a main breaker and a neutral bar. The neutral wire is wired through the meter with a jumper. The other side of the main breaker is called the *load side* and is wired to two electrical conductors in the panel that are called *buss*.

The voltage/current leaves the buss by way of a jumper through the branch breaker to the load (electrical consuming device) and returns to the panel by way of neutral if the voltage is 120 volts. If the voltage is 220 volts, single phase, the voltage will leave a branch breaker on line 1, go to the load, and return to the panel on the load side of a branch breaker off of line 2 buss. The voltage from line 1 to neutral is 120 volts. The voltage from line 2 to neutral is 120 volts. The voltage from line 1 to line 2 is 240 volts. The amps that the neutral carries on a three-wire circuit is the difference between the amps of line 1 and line 2.

The cold water pipe ground is wired to the neutral bar and, therefore, as long as the neutral is wired to cold water pipe ground the voltage of the neutral will be zero volts in respect to ground. The ground wire that connects to the load also connects to the cold water pipe ground at the panel. The schematic is a small electrical diagram of a power distribution system.

The purpose of the service head (Fig. 1-1) is to keep water from entering the service entrance conduit, meter, and main panel. The purpose of the service entrance conduit is to protect the service entrance conductors from physical damage and, in the case of fire, to protect against electrocution. Conduit is also used as a ground conductor. All branch wiring that leaves the main panel is normally in conduit. Some types of conduit are rigid (hard wall) conduit, flexible conduit, and thin wall conduit. Romex and knob and tube wiring do not use conduit. Romex is a wiring system with fiber or heavy plastic covering. Years ago, many buildings had open wiring within their walls that separated on the average of 2½ inches between conductors by knobs or tubes.

The step-down transformer shown in Fig. 1-1 is normally mounted on the power poles, is oil-filled, and air-cooled. The purpose of the step-down transformer is to take very high transmission voltage and step it down to a low voltage for the customers to purchase. As shown in Fig. 1-1, you will find the lower voltage to be on the lower cross members of the power poles. If there are three wires that leave the telephone poles and enter a home by way of service head and service entrance conduit, you will know that the voltage is 120/240 volts, single-phase, ac. The older homes built before 1940, many of them had only two wires; therefore, they were only 120 volts, single-phase. Only one line and the neutral of the secondary of the power transformer was used. In following the schematic we leave the power poles with wiring and feed the meter first. The meter is in series with the main breaker which is in series with all of the branch circuit breakers of the power panel. The branch circuit breaker is the last over current device that feeds an electrical load. In the case of 110 volts, we leave the branch breaker and feed the load and return to neutral. In the case of 220 volts we leave a branch breaker on line 1, feed a load, and return to a branch breaker on line 2.

Most of the wiring that is modified or changed will deal with branch circuit wiring or feeder wiring. A *feeder* is a wire that lies between a branch

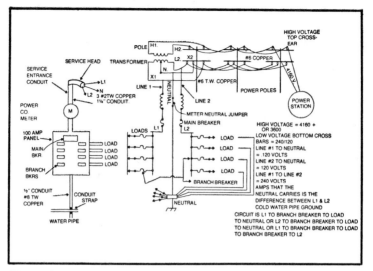

Fig. 1-1. A step-down transformer.

circuit breaker and a main panel breaker. You will find that branch circuit wiring can be many different sizes of wire as well as different numbers of wire in a conduit. The size of the wire is determined by the amperage consumption of the electrical load that you are feeding with that wire.

For example, if you have a machine that takes 20 amps of current, according to electrical code, you will find that you will need a No.12 T.W. wire. The number of the wire refers to the American wire gauge. The No. 12 wire will be copper wire. If we have a 30 amp load, we would need a No. 10 T.W. copper wire. A No. 8 copper wire equals 40 amps. The lower American wire gauge number means the diameter of the wire is larger and the current capacity of the wire is larger.

Aluminum wiring must be derated. This means you will have to use one size larger diameter wire to carry the same current as the copper wire. Example: a No. 10 T.W. aluminum wire is good only for 20 amps compared to a No. 12 copper, which is a smaller sized wire and is rated at 20 amps.

The diameter of conduit is dependent upon the size of wiring inside the conduit plus the number of wires in the conduit. The number of the wires inside of the conduit is known as *conductor fill*. The conductor fill of conduit is normally 30 percent to 40 percent of the cross sectional diameter of the conduit. You will find there are many types of *raceways* in the electrical field. A raceway is nothing more than an enclosure for the protection of wiring.

The amount of current a wire can carry is dependent upon the type of metal that the wire is made of (aluminum or copper), cross sectional diameter of the wire (American wire gauge), the type of insulation or covering for the wire, and whether the wire is suspended in free air or inside of electrical conduit. High-voltage wiring normally has a small current.

Therefore, the top cross members of voltage transmission lines have small wiring. Example: 4160 volts at 60 amps, which would be a No. 16 American wire gauge wire, would equal 249,600 watts. This figure was obtained by multiplying volts times the amps times the power factor (efficiency of the circuit).

On the higher voltage wire, the current was 60 amps. But let us see what happens when we are working with the lower voltage wiring. At 240 volts, the maximum watts that a No. 6 T.W. wire can handle if it is used inside of conduit will equal 14,400 watts. The current handled by the wire will remain the same regardless of voltage. But when we raise the voltage, you can see that the watts will also rise and the wire can handle more power at a higher voltage.

In looking at the step-down transformer, the watts of electricity that we supply the primary will equal the watts that leave on the secondary winding of the step-down transformer. The transformer has changed the voltage current relationship.

A note of interest is that the wiring that enters the house (service entrance conductors) will be larger than the transmission wiring that serves many buildings. A power transmission line equals high voltage, low current while service entrance conductors equal low voltage, high current. Watts remain the same in both cases with a high power factor. You must remember that you pay for watt-hours.

A 1-horsepower motor which is equal to 800 watts will cost the same to run at 120 volts as it does if you have the same 1-horsepower motor at 240 volts. The 800 watt-hours will be the same regardless of voltage.

Example:

$$\begin{array}{rl} & \text{Watts} = \text{volts} \times \text{amps} \\ 1\,\text{hp} & 800 = 120 \ \times \text{amps} \\ 120\,\text{volts} & 120 = 120 \\ & \dfrac{800}{120} = \text{amps} \\ & \text{amps} = 6\tfrac{2}{3} \end{array}$$

Example:

$$\begin{array}{rl} & 800 = 240 \times \text{amps} \\ 1\,\text{hp} & \dfrac{800}{240} = \dfrac{240 \times \text{amps}}{240 \times \text{amps}} \\ 240\,\text{volts} & \\ & \text{amps} = 3\tfrac{1}{3} \end{array}$$

120 volts x 6⅔ amps = 240 volts x 3⅓ amps

800 watts = 800 watts

You pay for watt-hours (watts **x** time in hours). Many salesmen will attempt to sell a higher voltage machine at the same horsepower with the understanding that the current is less and because of this the running cost will be less. This whole line of reasoning is false.

Figure 1-2 shows a three-phase power distribution panel and meter. Line 1, line 2, and line 3, all have hot voltage when energized. Voltage can be 208, 240, 480, or higher. Note the delta wired load on the left of the drawing and the star wired load on the right of the drawing. All three phase loads are either wired delta or star.

The symbols θ = phase which means time and $3\,\theta$ electricity is composed of three lines each spaced 120° time of voltage or ampere sine wave cycle apart. $3\,\theta$ has three sine waves of voltage and three sine waves of amperes per each cycle.

The ampere and voltage sine waves of a resistance circuit will be in step with each other. Normally, they are different wave heights, but the

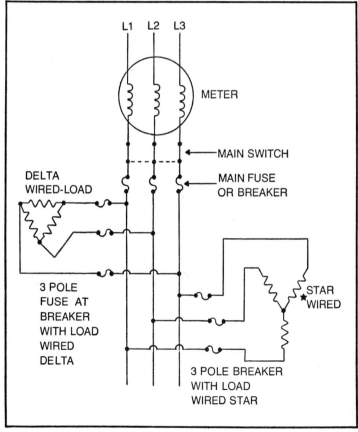

Fig. 1-2. A three-phase power distribution panel.

voltage wave intersects the amperage wave at 0, 180, and 360° of phase time. A coil (inductance) will lower the power factor. Voltage sine wave will lead amperage sine wave.

A capacitor will lower the power factor. Amperage sine wave will lead voltage sine wave. An inductance circuit with a lower power factor can have a capacitor placed in parallel. With proper balance, the power factor can be corrected above .9. This process can be reversed. A low power factor means the efficiency of the circuit is bad and you end up paying for power that does no economic work.

ELECTRICITY

Electricity is a form of energy that is a result of electrons in motion. Electricity has three properties. These are voltage, current, and resistance. Voltage is electromotive force that pushes a flow of electrons known as current, through a conductor or medium that has a resistance to slow down the atoms that are being pushed.

Ohm's law states that it requires an electromotive force of 1 volt to push a current of 1 ampere through a resistance of 1 ohm. *Voltage* is electromotive force, potential, or electrical pressure. *Current* is the flow of electrons that is being pushed along a conductor. *Resistance* is the ability of a substance or conductor to impede the flow of electrons that are being pushed by voltage. *Power* is the capacity of electricity to do work. It is equal to the volts times the amps of a circuit and is expressed in watts. Watts is equal to power which is equal to volts times amps. The letter symbols are as follows:

E stands for voltage, electromotive force, or electrical pressure.
I stands for current.
R stands for resistance = Ω = ohms as a unit of measure.
Watts is equal to power = E x I = volts times amps.

OHM'S LAW

A triangle is shown in Fig. 1-3 to illustrate Ohm's law. Use this triangle to help solve problems when you have any two of the components in the triangle.

Step No. 1. Cover up with your finger the component that you want to solve and the relationship will be plainly visible to you. Example: If you want to solve for E (voltage), cover up E with your finger.

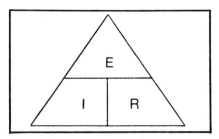

Fig. 1-3. Ohm's law triangle.

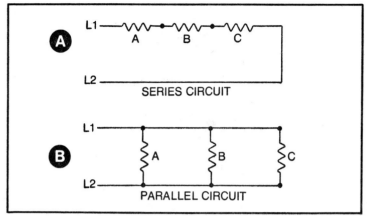

Fig. 1-4. Circuits.

Step No. 2. You will see I x R is visible to you and that is the solution.

If you want to solve for I (current), Step No. 1 is to cover I with your

finger. Step No. 2 you will see the relationship of $\frac{E}{R}$ and that is your solution.

If you want to solve for R (resistance), cover R with your finger. Step

No. 2 you will see the relationship of $\frac{E}{I}$ and that is your solution.

A series circuit (Fig.1-4A) is where one electrical device follows another electrical device. An example is a switch wired in series with the load it controls. A parallel circuit (Fig.1-4B) is where one electrical device is beside another electrical device.

With parallel resistance problems (Fig. 1-5) you will find that the resistance total will be equal to a value which is smaller than the smallest

resistor in the problem. The resistance total of $\frac{10}{13}$ ohms is less than the

smallest resistance R_3 which was 1 ohm.

Watts is equal to volts times amps = E x I. (E equals volts; I equals current which is amps).

For an air conditioner rated at 1200 watts with voltage at 120 volts, what is the current or amps? Example:

$$\text{Watts} = \text{volts times amps}$$
$$\frac{1,200}{1,200} = \frac{120}{120} \text{ x amps}$$
$$\frac{1,200}{120} = \frac{120}{120} \text{x amps}$$
$$\frac{1,200}{120} = \frac{120}{120} \text{ x amps}$$
$$\text{Amps} = 10$$

$$\text{MILLI} = \frac{1}{1,000} = .001 \text{ TIMES UNITS OF MEASURE.}$$
EXAMPLE: MILLIVOLTS, MILLIAMPS, MILLIWATTS.

$$\text{MICRO} = \frac{1}{1,000,000} = .000001 \text{ TIMES UNIT OF MEASURE.}$$
EXAMPLE: MICROVOLTS, MICROAMPS, MICROWATTS.

KILO = 1,000 TIMES THE UNIT OF MEASURE.
EXAMPLE: KILOVOLTS, KILOWATTS

MEGA = 1,000,000 TIMES THE UNIT OF MEASURE.
EXAMPLE: MEGAWATTS OR MEGOHMS.

RESISTANCE SERIES FORMULA: $R_t = R_1 + R_2 + R_3$

EXAMPLE: $R_1 = 10$ OHMS
$R_2 = 5$ OHMS
$R_3 = 1$ OHM

$R_t = 10 + 5 + 1 = 16$ OHMS

(R_t STANDS FOR RESISTANCE TOTAL).

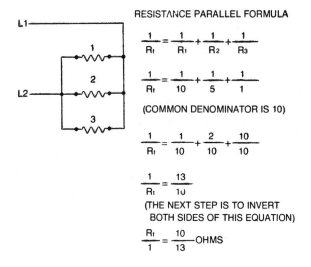

RESISTANCE PARALLEL FORMULA

$$\frac{1}{R_t} = \frac{1}{R_1} + \frac{1}{R_2} + \frac{1}{R_3}$$

$$\frac{1}{R_t} = \frac{1}{10} + \frac{1}{5} + \frac{1}{1}$$

(COMMON DENOMINATOR IS 10)

$$\frac{1}{R_t} = \frac{1}{10} + \frac{2}{10} + \frac{10}{10}$$

$$\frac{1}{R_t} = \frac{13}{10}$$

(THE NEXT STEP IS TO INVERT
BOTH SIDES OF THIS EQUATION)

$$\frac{R_t}{1} = \frac{10}{13} \text{ OHMS}$$

Fig. 1-5. Resistance problems.

All general building wiring is rated at 600 volts ac. The current depends on wire size (diameter of the wire, American wire gauge), type of material (copper, aluminum), and type of insulation. Wiring is sized to the current consumption of the load. See Table 1-1 and 1-2.

Wiring is sized to conduit. The conduit is a metal enclosure that protects the wiring and gives you a ground return for safety. Exceptions are nonmetallic conduit (plastic PVC), cords, and Romex (a fibrous covered wiring system).

The ground on nonmetallic conduit systems is accomplished by an additional green-colored ground wire. In the metal conduit systems, the ground is accomplished by the metal of the conduit and the conduit is the

Table 1-1. Ampacity Copper Conductor.

SIZE	NAME* TW	NAME RH, RHW, THW, THHN.	NAME AVA	NAME A
14	15	15	30	30
12	20	20	35	40
10	30	30	45	55
8	40	45	60	75
6	55	65	80	95
4	70	85	105	120
2	95	115	135	165
1	110	130	160	190
0	125	150	190	225
00	145	175	215	250
000	165	200	245	285
0000	195	230	275	340
250	215	255	315	----
300	240	285	345	----
350	260	310	390	----
400	280	335	420	----
500	320	380	470	----
600	355	420	525	----
700	385	460	560	----
750	400	475	580	----
800	410	490	600	----
900	435	520	----	----
1000	455	545	680	----
1250	495	590	----	----
1500	520	625	785	----
1750	545	650	----	----
2000	560	665	840	----

*Name codes are given in Table 1-5

Table 1-2. Ampacity Aluminum Conductor.

SIZE	NAME* TW	NAME RH, RHW THW, THW- N	NAME AVA	NAME A
12	15	15	25	30
10	25	25	35	45
8	30	40	45	55
6	40	50	60	75
4	55	65	80	95
2	75	90	105	130
1	85	100	125	150
0	100	120	150	180
00	115	135	170	200
000	130	155	195	225
0000	155	180	215	270
250	170	205	250	----
300	190	230	275	----
350	210	250	310	----
400	225	270	335	----
500	260	310	380	----
600	285	340	425	----
700	310	375	455	----
750	320	385	470	----
800	330	395	485	----
900	355	425	----	----
1000	375	445	560	----
1250	405	485	----	----
1500	435	520	650	----
1750	455	545	----	----
2000	470	560	705	----

*Name codes are given in Table 1-5

ground conductor. The conductor fill inside of the conduit is an average of 30 percent to 40 percent of the internal diameter area of the conduit.

Many small compressor condenser sections are wired by cord instead of conduit. The cords are rated at 250 volts or 600 volts. The cords are coded S.O., S.J.O., S.J.

S.O. cord is heavy duty, thick rubber-walled, and the inside has jute filler. S.J.O. cord is medium duty, medium rubber-walled thickness, and the inside has paper and jute filler. S.J. cord is junior service duty, thin rubber-walled thickness, and the inside has paper filler. See Table 1-3 for amperage.

Here are the steps to take in figuring out the amperage of a 1600-watt compressor section at

120 volts	220 volts
Watts $= E$ times I	Watts $= E$ times I
$1600 = 120 . I$	$1600 = 220 . I$
$\dfrac{1600}{120} = \dfrac{120 . I}{120}$	$\dfrac{1600}{220} = \dfrac{220 . I}{220}$

16

Table 1-3. Cord Table

SIZE	NAME* S, SO, SJ, SJO, THERMO- PLASTIC	WATTS VOLTAGE 120V/240V
18	7 AMPS	840/1180
16	10	1200/2400
14	15	1800/3600
12	20	2400/4800
10	25	3000/6000
8	35	4200/8400
6	45	5400/10800
4	60	7200/14400
2	80	9600/19200
CORD RATED AT 250 or 600 VOLTS ac SEE CORD LABEL. SMALL COMPRESSOR CONDENSER UNITS USE CORD AND ARE RATED IN AMPS OR WATTAGE		
*Name codes are given in Table 1-5		

Divide both sides of the equation by 120.

$$\frac{\overset{13.3}{1600}}{120} = \frac{120 \cdot I}{120}$$

$$I = 13.3 \text{ amps}$$

Divide both sides of the equation by 220.

$$\frac{\overset{7.27}{1600}}{220} = \frac{220 \cdot I}{220}$$

$$I = 7.27 \text{ amps}$$

The cord or wire for the 120-volt compressor condenser section will be No. 14 SO or No. 14 SJO cord. The conduit and wire will be two No. 14 T.W. wires in one-half inch conduit (copper). The cord or wire for the 240-volt compressor condenser section will be No. 16 SO or No. 16 SJO cord. The conduit and wire will be two No. 14 TW wires in one-half inch conduit (copper). You cannot use a smaller wire on this application because No. 14 is the smallest wire allowed for general building wiring. If you use aluminum wire, you would use two No. 12 TW aluminum wires in one-half inch conduit for the 120-volt section and 240-volt section. Aluminum wire is derated so you have to use a larger size and the No. 12 aluminum will handle either section. See Tables 1-4 and 1-5.

Table 1-4. Conduit and Wire.

BASED UPON 40% CONDUCTOR FILL WITHIN THE INSIDE DIAMETER OF CONDUIT

WIRE PHYSICAL IDENTIFICATION

WIRE SIZE	WIRES IN WIRE BODY	WIRE DIAMETER INCHES	WIRE NAME	WIRE SIZE	1/2	3/4	1	1¼	1½	2	2½	3	3½	4
18	1	.0403	TW											
16	1	.0508	THW											
14	1	.0641	THHN	14	9	15	25	44						
12	1	.0808		12	7	12	19	35						
10	1	.1019		10	5	9	15	26						
8	1	.1285		8	3	5	8	14						
6	7	.0612		6	1	2	4	7						
4	7	.0772		4	1	1	2	5						
2	7	.0974		2	1	1	3	4	4	4				
1	19	.0664		1			1	3	3	3	4	4		
0	19	.0745		0			1	2	2	2	3	3		
00	19	.0837		00			1	1	1	2	3	3	4	
000	19	.0940		000			1	1	1	1	2	2	4	4
0000	19	.1055		0000			1	1	1	1	1	2	3	4
250	37	.0822		250			1	1	1	1	1		3	
300	37	.0900		300				1	1	1	1			
350	37	.0973		350				1	1	1				
400	37	.1040		400				1	1					
500	37	.1162		500					1					
600	61	.0992		600										
700	61	.1071		700										
750	61	.1109		750										
800	61	.1145												
900	61	.1215												
1000	61	.1280												
1250	91	.1172												
1500	91	.1284												
1750	127	.1174												
2000	127	.1255												

BY USING A MICROMETER ON EACH WIRE STRAND IN THE WIRE BODY, YOU WILL BE ABLE TO IDENTIFY WIRE SIZE. YOU MUST KNOW WIRE SIZE TO DETERMINE IF THE CIRCUIT IS OVERLOADED.

Table 1-5. Wire Guide.

NAME	DESCRIPTION	MAX. TEMP.	USAGE
RH	RUBBER COVERED WITH FIBER. HEAT RESISTANT	167 F.	DRY PLACES. NOT USED VERY MUCH. NEED TO SPECIAL ORDER.
RHH	RUBBER COVERED WITH FIBER. HEAT RESISTANT	194 F.	DRY PLACES. NOT USED VERY MUCH. NEED TO SPECIAL ORDER.
RHW	RUBBER COVERED, HEAT AND MOISTURE RESISTANT.	167 F	DRY AND WET PLACES. NOT USED VERY MUCH. HARD TO STRIP AND FISH.
T	THERMOPLASTIC	140 F	DRY PLACES
TW	THERMOPLASTIC, MOISTURE RESISTANT	140 F	DRY AND WET PLACES. MOST COMMON WIRE.
THHN	THERMOPLASTIC, HEAT RESISTANT	194 F	DRY PLACES. SKINNY WIRE. USE REWORK.
THW	THERMOPLASTIC, HEAT AND MOISTURE RESISTANT.	167 F	DRY AND WET PLACES
THWN	THERMOPLASTIC, HEAT AND MOISTURE RESISTANT.	167 F	DRY AND WET PLACES. SKINNY WIRE. USED IN REWORKING OLD WORK. MORE CIRCUITS
A	ASBESTOS	329 F	DRY PLACES. STRIP HEATERS. HIGH TEMP.
V	VARNISHED CAMBRIC	185 F	DRY PLACES. MOTORS
AVA	ASBESTOS AND VARNISHED CAMBRIC	230 F	DRY PLACES. STRIP HEATERS. HIGH TEMP.
AVL	ASBESTOS AND VARNISHED CAMBRIC	230 F	DRY PLACES. STRIP HEATERS. HIGH TEMP.
S & SO CORD	RUBBER JACKET WITH PAPER AND JUTE FILLER	SEE LABEL	HEAVY DUTY. DRY AND WET PLACES. MARINE WORK. SMALL COMP. COND. SECTIONS
SJO & SJ CORD	RUBBER JACKET WITH PAPER FILLER	SEE LABEL	LIGHT DUTY , DRY AND WET PLACES. PORTABLE WORK.

ALL GENERAL BUILDING WIRING RATED AT 600 VOLTS
S, SO, SJO, SJ CORD RATED AT 250 or 600 VOLTS. SEE CORD LABEL FOR OTHER INFO.

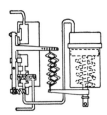

Chapter 2
Electric Meter
Construction & Theory

In meter applications, the same meter movement can be used for the voltmeter, ohmmeter, or ammeter. The only difference is that the components are arranged in a different order. Figure 2-1 is a drawing of a voltmeter. The resistor is in series with the meter movement. Because the meter movement can only handle a maximum of 3 volts, the resistor is a drop resistor which will drop the excess voltage and you will read a voltage which is proportional to the applied voltage of the circuit you are attempting to measure.

As in Fig. 2-1, drop 477 volts across resistor A. The meter movement will swing to full scale. This is because you are using the full 3 volts. When you see the components wired as in Fig. 2-1, you will always have a voltmeter.

CHECKING VOLTAGE

■ Turn off power on the load.

■ Place the meter in parallel with the load. Start with the highest voltage range.

■ Turn on the power and work your way down the voltage ranges until you read near the center of the meter. Multiply the meter reading times the voltage range and this is the true voltage. The voltmeter on its highest range can not be damaged because of the high resistance of its circuit.

Figure 2-2 shows an ohmmeter and its components. With an ohmmeter, we place a small dc voltage through a load and read the dropped voltage on the meter. This dropped voltage is equal in ohms resistance dc.

USING THE OHMMETER

■ Make sure there is no line voltage on the load circuit.

■ Place the meter in parallel with the load.

■ Start with the highest resistance range and work down to where you read in the center of the meter, if possible. Check to see if you have

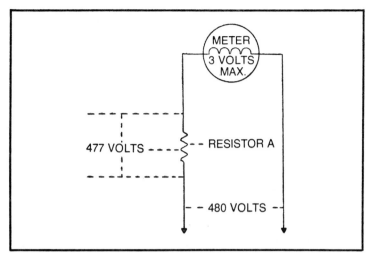

Fig. 2-1. The meter movement will indicate full scale.

continuity to the load ground, case or shell. When the load you are checking is of low ohms, most of the battery voltage is dropped across the meter drop resistor.

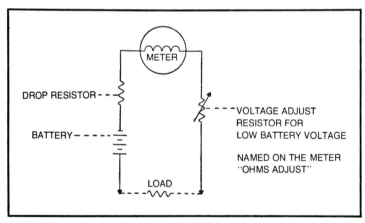

Fig. 2-2. An ohmmeter and its components.

When the load resistance is high, most of the battery voltage is dropped across the load. If you have continuity to the load ground, case or shell, you have a short circuit.

There must be no line voltage connected to the meter. If there is line voltage on the meter it will burn up the meter. The ohms of the meter circuit is too low.

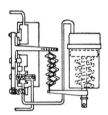

Chapter 3
Refrigeration
Compressor Continuity

The *hermetic compressor* is a compressor that is totally sealed. There are no nuts and bolts on this compressor and it cannot be dismantled with wrenches. Larger sizes of hermetic compressors can be taken apart by grinding the bead weld apart and separating the compressor dome from the crank case and removing the inside components.

The *semihermetic compressor* is a compressor that is held together by nuts and bolts and can be taken apart by wrenches and overhauled with new reed valves, pistons, rods and bearings. The motor windings can be removed and rewound.

The *open drive compressor* is a refrigeration compressor that is of nut and bolt construction with a shaft that extends out of the case with a shaft seal. This shaft can be coupled to an electric motor, gasoline or diesel engine, or steam turbine if it is a large capacity compressor.

Reciprocating compressors use valves and pistons; we have centrifical compressors that are constructed similarly to centrifical water pumps and there are no valves; and there are rotary compressors.

With hermetic and semihermetic compressor motors you will find two kinds or motor terminals. The first motor terminal is where you have pins that extend out of Bakelite. The second kind of motor terminal is a nut and bolt motor terminal.

The nut and bolt terminals were used years ago on small fractional horsepower compressor motors. Usage was discontinued because of refrigerant leak problems. The nut and bolt motor terminal is still very common in use on large horsepower compressor motors. When you are going into a compressor ring out (continuity checkout) the first thing you must do is to draw out the pins or nut and bolt motor terminals as they appear in front of you.

Normally, if you have pins and Bakelite for your motor terminals, you will find that the pins will be in a triangle configuration extending through the round Bakelite seal. The nut and bolt motor terminals are normally in a straight line or in a line layout. Take a piece of paper and draw the pins as they actually appear on the compressor motor case. See Fig. 3-1.

The first thing you have to do in this continuity checkout is to secure an ohmmeter that is able to distinquish 1 ohm from 10 ohms. The inexpensive meters will not be able to distinquish 1 ohm from 10 ohms and you will have to purchase a $25 to $30 meter to distinquish these low resistances.

When you are ringing out the windings of a compressor you are dealing with 1, 5, 10, 15, or 20 ohms resistance and you must be able to distinquish between these ohmages. Take the ohmmeter and zero in the lowest ohm scale with the zero "adjust" knob and take readings across the pins in the configuration. After each reading, write in the ohm value across the pins from which you took the reading. Note example A and example B of Fig. 3-2 to see how the pins and the ohmage resistance have been drawn. Example A is a triangular configuration and example B is an in line configuration.

■ The highest reading between the pins either in the triangle configuration or the in line configuration tells you that you are across the start and run pins.

■ The pin that is remaining becomes common (process of elimination).

■ After you have located the common pin, line out or cover over the highest reading. This reading is only needed to establish the start and run pins and to locate the common pin that was remaining.

■ The lowest reading from the common pin is the run winding. Normally, it is about 1 ohm. The highest reading from the common pin is the start winding. Normally , it is about 5 to 22 ohms.

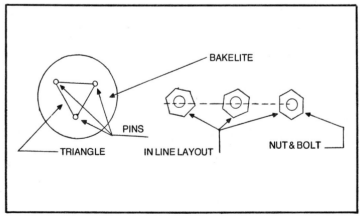

Fig. 3-1. Continuity checkout.

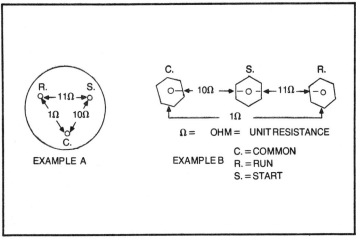

Fig. 3-2. A triangular and a line configuration.

■ If you find all three readings have the same ohms, this compressor is 3-phase. There are 3 run or 3 main windings.

■ The rotation is critical if the compressor is centrifical or rotary 3-phase. For all other compressors the rotation is not important. The compressor will raise pressure, move refrigerant and the oil pump of the compressor will pump oil in either direction. Rotation can be changed by interchanging any two line leads at breaker, starter, disconnect switch or motor terminals. Note example A of Fig. 3-3. Example B of Fig. 3-3 shows nut and bolt layout.

If the run winding is less than one-half ohm or the start winding is less than 3 ohms, you have a short circuit in the winding. The compressor is no good. If there is continuity of the start or run winding to the compressor case, you have a short circuit and the compressor is no good. If there is only one reading across 2 pins, there is an open winding or burned out winding and the compressor is no good. See Fig. 3-4.

LOCKED ROTOR TEST

■ Turn off the power to the compressor.

■ Remove all machine wiring attached to the compressor motor.

■ Ring out the compressor and label the common, start, and run pins.

■ Secure line 1 to the run pin and line 2 to the common pin.

■ Place the jumper wire insulated from run to start. (See A of Fig. 3-5).

■ Turn on the power.

■ Compressor will start and come up to rpm in less than 5 seconds.

■ After compressor has come up to rpm, remove jumper wire with the compressor running, line 1 to run, and line 2 to common.

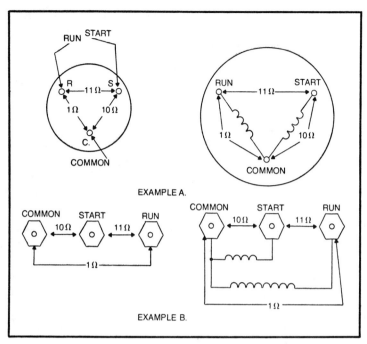

Fig. 3-3. Rotation.

To start the Capacitor Start Hermetic Compressor Motor, repeat steps one through four for the induction start, induction run motor. Step five is to place jumper wires from run to start capacitor and from start to start capacitor. It makes no difference on the hook-up of the jumper wires to the start capacitor terminals.

Repeat steps six and eight for the induction start, induction run motor. On step eight, remove jumper wires and start the capacitor with the compressor running. Line No. 1 to run and line No. 2 to common.

Round off the tenths and use the running current information in Table 3-1 to tell you if the compressor is drawing too much current per horsepower. You can also use Tables 3-1, 3-2 and 3-3 to size klixons and heaters for magnetic starters if you do not know the running current of the compressor. When charging a refrigeration system, do not add refrigerant when the current reaches these values. You will be overcharging the system. Use Table 3-3 to check for locked rotors. To roughly estimate the horsepower of a compressor, these tables can be used in reverse.

LOCKED ROTOR REPAIR

Normally, the compressor will try to start but it will hum and the klixon will open or the fuse or breaker will blow. The klixon is a switch that is mounted on the side of the compressor, looks like a miniature hat, and will open with excess currents or excess compressor case temperature. The klixon will reset when it becomes cool. You should be able to touch by hand all motor cases for at least 5 seconds. If the motor case is so hot that you cannot touch it for 5 seconds, there is something wrong.

If you believe you have a locked rotor, remove wiring from the compressor motor terminals and use the line cord and jumper method described in the section on *refrigeration compressor run out.* If all you hear is a hum, you have a locked rotor. Make sure you have done the continuity check. You could have open winding or a short. Also, check out the run and start capacitor (see Chapter 4).

HOT-SHOTTING THE COMPRESSOR

Step No. 1: Make sure the start and run capacitors have high enough voltage ac rating for the new applied voltage.

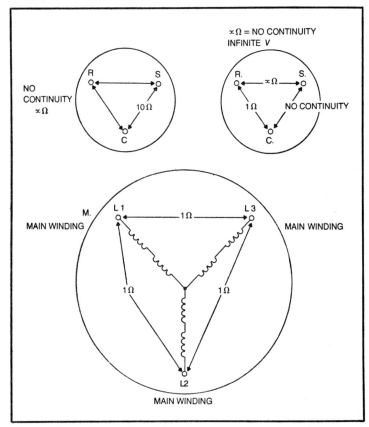

Fig. 3-4. Three-phase compressor windings have three readings. They all have the same ohmages with no continuity to the compressor shell.

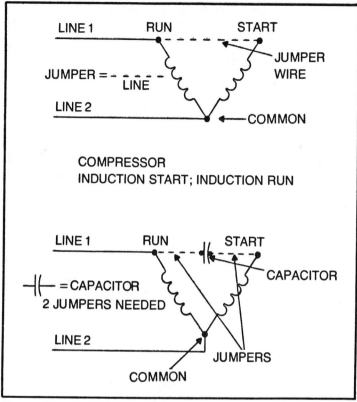

Fig. 3-5. A capacitor start hermetic compressor motor diagram.

Step No. 2: Remove wiring from the compressor motor terminals. Double the line voltage and hookup as in Fig. 3-6.

Step No. 3: Make sure the power is off while you are doing the second step.

Step No. 4: If the compressor is 120 volts ac, make the line No. 1 to No. 2 240 volts ac. If the compressor is 240 volts ac, single phase, make line No. 1 to No. 2   480 volts ac, single phase.

Step No. 5: Attach a jumper from run to capacitor and a second jumper from start to capacitor.

Step No. 6:   Take the jumper wire off the start terminal of the compressor.

Step No. 7: Turn on the higher voltage.

Step No. 8:   Take the jumper wire and tap it about four times (one second each to the start terminal). Do not touch the live voltage. Hold on to the insulation of the jumper wire.

Step No. 9: Turn off the power and repeat the above procedure in 5 minutes.

Table 3-1. Running Current (Three-Phase).

AC-MOTORS FULL LOAD.			
HP	240 V	480 V	2300V
½	1.91	.96	
¾	2.68	1.34	
1	3.45	1.73	
1½	4.98	2.49	
2	6.51	3.26	
3	9.19	4.6	
5	14.55	7.28	
7½	21.05	10.53	
10	26.8	13.4	
15	40.2	20.1	
20	51.68	25.84	
25	65.08	32.54	
30	76.56	38.28	
40	99.53	49.77	
50'*	124.41	62.21	
75	183.74	91.87	20
100	237.34	118.67	26
60	147.38	73.69	16

REVERSE ROTATION METHOD

Step No.1: Turn off power and remove the wiring from Hermetic compressor motor terminals.

Step No. 2: Study Fig. 3-7.

Step No. 3: Wire line 1 to common and line 2 to start. Use compressor rated line voltage.

Step No. 4: Attach the jumper wire from run to capacitor and second jumper wire from start to capacitor.

Step No. 5: Disconnect jumper from run motor terminal.

Table 3-2. Running Current (Single-Phase, Full Load ac Motor).

HP	120	240
1/6	4.22	2.11
1/4	5.56	2.78
1/3	6.9	3.45
1/2	9.39	4.69
3/4	13.22	6.61
1	15.33	7.66
1 1/2	19.16	9.58
2	23	11.5
3	32.57	16.29
5	53.65	26.83
7 1/2	76.64	38.32
10	95.81	47.9

Table 3-3. Average Motor Locked Rotor Amps.

HP	SINGLE PHASE		THREE PHASE	
	120 V	240 V	240 V	480 V
½	56.33	28.17	11.49	5.75
¾	79.32	39.66	16.08	8.04
1	91.97	45.98	20.1	10.05
1½	114.96	57.48	28.71	14.36
2	137.95	68.98	37.32	18.66
3	195.43	97.91	51.68	25.84
5	421.89	160.94	86.13	43.07
7½	459.84	229.92	126.32	63.16
10	574.8	287.4	155.08	77.54
15			229.68	114.84
20			298.58	149.29
25			367.49	183.75
30			447.88	223.94
40			597.17	298.59
50			717.75	358.88
60			861.3	430.65
75			1062.27	531.14
100			1412.53	706.27
125			1780	890

Step No. 6: Turn on the line voltage and hold jumper wire by the insulation; then hold jumper to run for four seconds, four times at four second intervals.

Step No. 7: If the motor does not reverse, repeat step six using 240 volts instead of 120 volts motor rated voltage. Use 480 volts instead of 240 volts motor rated voltage (single phase).

Check the capacitor for a higher rated voltage. When you have completed steps one through seven using double the line voltage, you have hot-shotted the compressor in reverse. Do not hot shot in reverse until you have done steps one through seven with compressor rated line voltage. You might be able to break the locked rotor with normal line voltage.

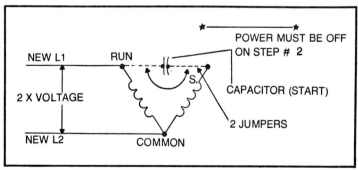

Fig. 3-6. Double the line voltage and hookup.

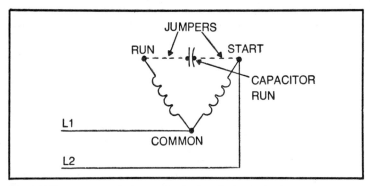

Fig. 3-7. The reverse rotation method to break a locked rotor.

There is no need to strain the compressor motor unless necessary. If these steps do not free the locked rotor, there is nothing else you can do in the field.

REFRIGERATION VALVES (COMPRESSOR REED VALVES)

When the compressor is in the refrigeration system and is running, if the suction and the high-side pressures are the same, the valves are bad. With the compressor out of the system and running, take a finger and hold it over the compressor suction pipe (large pipe). If the finger is sucked in and held sucked in, there is good chance the suction valves are good. With the compressor running, hold your finger over the discharge pipe (small pipe). After 5 seconds, remove your finger and see if you can hear, the

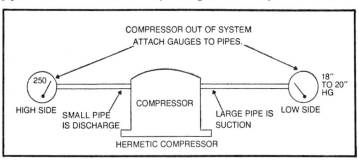

Fig. 3-8. Check the pressures.

swish of air pressure. If you do, there is a good chance the discharge valves are good. If you have gauges, check the pressures. See Fig. 3-8.

Only one problem is the diagnosis of weak valves. The compressor needs to be in the refrigeration system. Start up the system and run the compressor 10 minutes or more. Insert gauges on suction and discharge. Observe the pressures. Shut the unit off. If pressures equalize within 20 seconds or less, and the sweat or frost line goes away within 20 seconds, you have weak valves.

Chapter 4
Checking Capacitors

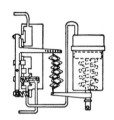

The start capacitor has a range of 95-400 microfarads. It resembles a cylinder and it usually has a black plastic case. The run capacitor has a range of 5-40 microfarads. It has a metal case and it is physically larger than the start capacitor. Many times there will be two run capacitors wired in parallel with a bleeder resistor across their terminals as a safety device. The run capacitor is built quite heavy with oil, tar, or waxed paper. As a dielectric, it must be rated continuous duty. Continuous duty means that the electrical device is in service for more than 4 hours per day. All refrigeration, air conditioning, and heating units are classed as continuous duty.

Step No. 1. In checking capacitors, the first step is to take your voltmeter, ohmmeter, ammeter, or multimeter and set it up on the selector switch for highest resistance ohms scale. See B of Fig. 4-1. A of Fig. 4-2 represents the inside components of a small multimeter set up for ohms. Observe that No. 1 is a battery, No. 2 is a voltage drop resistor, No. 3 is the meter movement, and No. 4 is a variable resistor for ohms adjust to zero in the ohmmeter.

All four components are wired in series. By selecting the highest ohm scale, we are placing a resistor of lowest resistance for No. 2. On the large microfarad capacitors, I want to utilize the maximum voltage/current capacity of the meter battery to place a charge on the plates of the start capacitor shown in B of Fig. 4-1.

Step No. 2. Short the prods of the meter together and adjust the knob to make the needle point to zero ohms. Dead short or no resistance.

Step No. 3. Take a jumper and short out the charge that might be on the plates of the capacitor. Remove the shunt resistor if one is wired across the capacitor. This step is very important as you could get a shock or damage the meter movement.

PLACING A CHARGE TEST

Step No. 4. After the capacitor has been discharged, remove the jumper and place the meter prods on the capacitor so as to place dc charge on the plates of the capacitor. If the meter needle swings to direction A and returns to direction B within 10 seconds, the start capacitor has taken the charge on the plates and the first half of the test is successful. The capacitor is no good if the needle goes to the low ohm reading and remains there never to swing back to infinite ohms. See B of Fig. 4-1. The capacitor has an internal short.

If the capacitor has a metal case, you must check continuity of the plates of the capacitor to its case. If there is an ohm resistance reading, the capacitor is no good. When you do step No. 4 and nothing happens on your highest resistance scale, the capacitor is open and no good. Presume step No. 4 is completed with no problems.

HOLDING A CHARGE TEST

Step No. 5. Set up the meter on the 10-volt scale dc or voltage near this. Figure 4-2 is a voltmeter as set up in step No. 5. A voltmeter will measure the excess voltage over the drop resistor and this will relate to a real voltage. Figure 4-3 shows the meter setting for voltage (10 volts dc) The second half of the test will determine if the capacitor held the charge that you put on its plates on step No. 4.

Step No. 6. Touch the prods of the meter to the capacitor. If the meter needle wants to go off the scale from zero volts to the left, quickly remove prods and reverse them on their hook up to the capacitor.

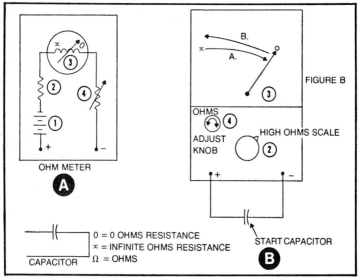

Fig. 4-1. Set the selector switch for the highest resistance.

Fig. 4-2. Because most meter movements can only handle a maximum of 2 volts, the drop resistor is needed to take care of the excess voltage. Number 1 is a drop resistor and number 2 is the meter movement.

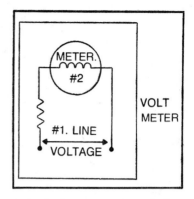

Step No. 7. When the prod's polarity is the same as the capacitor plate polarity, the needle of the meter will swing to the 10-volts mark or swing to A direction and return in B direction with the voltage charge on the plates of the capacitor being discharged through the No. 1 drop resistor and No. 2 meter movement. Step No. 7 tells you that the capacitor took a charge, held a charge and that the charge is good.

If nothing happens on No. 7 step, the capacitor is no good. If the meter just barely moves, the capacitor can be presumed to be weak and no good. In the case of small capacitors, as in radios and television sets, there will be little movement and the capacitor will be good. I like a good swing of the meter needle on placing the charge test and holding a charge test.

Checking a run capacitor is the same as the start capacitor with one exception. Step No. 1 will be set for the lowest resistance scale. In doing this, you are placing resistor No. 2 with high ohms. You don't need as much volt/ampere charge. However, no harm will be done to the capacitor if you use the high scale. Through experience, you will pick the scales of the meter to give full needle swing. You will want full needle swing on placing a charge test and holding a charge test.

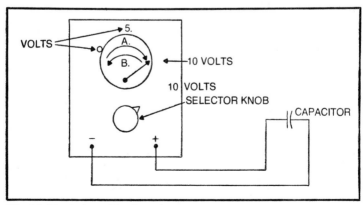

Fig. 4-3. The meter setting for voltage.

Table 4-1. Start Capacitor.

COMPRESSOR 1/8 HORSEPOWER	CAPACITOR WILL BE NEAR 95 TO 200mF
COMPRESSOR 1/6 HORSEPOWER	CAPACITOR WILL BE NEAR 95 TO 200mF
COMPRESSOR 1/4 HORSEPOWER	CAPACITOR WILL BE NEAR 200 TO 300mF
COMPRESSOR 1/3 HORSEPOWER	CAPACITOR WILL BE NEAR 250 TO 350mF
COMPRESSOR 1/2 HORSEPOWER	CAPACITOR WILL BE NEAR 300 TO 400mF
COMPRESSOR 3/4 HORSEPOWER	CAPACITOR WILL BE NEAR 300 TO 400mF

WHAT TO DO IF YOU CAN'T READ THE CAPACITOR

Suppose time and wear have erased the identification of the capacitor. Select a working voltage at the line voltage or above line voltage. If the case of the capacitor is plastic or black Bakelite you have a start capacitor and their range is 95 to 400 microfarads.

Three-phase compressors and motors do not have start or run capacitors. Start and run capacitors are wired between start and run winding. Start and run capacitors are in parallel at motor start up. The start capacitors drop out of the circuit when the motor is at about one-half of its rated rpm. A starting relay that is sensitive to voltage (potential) or current is used to switch out the starting capacitor because a set of contact points cannot be used inside of a hermetic compressor motor. The oil and refrigerant would make them useless in a short time. The run capacitor remains in the circuit at all times. A motor that has only a run capacitor is known as a permanent split capacitor type motor.

Consult Table 4-1 and 4-2. If the capacitor is inserted in the circuit and the compressor does not start, stop and select the next larger size capacitor. If the compressor does not start, but makes a hum noise, select a larger size again. When you have started the compressor and have it running—and you have not developed a hum while it is running—you are close enough to get by. However, if a hum develops while the compressor is running, you have gone too high in microfarads. Stop the compressor and wire in a smaller size capacitor. Tables 4-1 and 4-2 can be used as a guide and might have to be amended by experience.

Note that the start capacitor is only in the circuit at the instant of start until the motor has reached about 50 percent of its rpm and will not be in

Table 4-2. Run Capacitor.

COMPRESSOR or motor		CAPACITOR WILL BE NEAR
COMPRESSOR or motor	1/8 HORSEPOWER	CAPACITOR WILL BE NEAR 4 to 5mF
COMPRESSOR or motor	1/6 HORSEPOWER	CAPACITOR WILL BE NEAR 4 to 5mF
COMPRESSOR or motor	1/2 HORSEPOWER	CAPACITOR WILL BE NEAR 10
COMPRESSOR or motor	1/2 to 2 HORSEPOWER	CAPACITOR WILL BE NEAR 10 to 15mF
COMPRESSOR or motor	3 HORSEPOWER	COMPACITOR WILL BE NEAR 2 capacitors at 10mF. Total 20mF
COMPRESSOR or motor not to exceed 40 microfarads.	5 HORSEPOWER	CAPACITOR WILL BE NEAR 2 capacitors

the circuit until the next time the motor compressor is started. You can be a little off on the value, but as long as the motor compressor starts under load, that is all you need.

If the running currents of the motor compressor are printed on the case, a clamp-on ammeter could help in the selection of the run capacitor. Many times the only current the manufacturer will show is locked rotor amps. This is why I have developed a table of running currents for 3-phase motors, a table of running currents for single-phase motors, and an average motor locked rotor ampere table for single-phase and 3-phase motors. See Tables 3-1, 3-2, and 3-3.

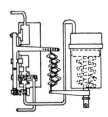

Chapter 5
Wiring & Controls

All single-phase motors and compressor motors have two windings except shaded pole (one winding) and GE compressor motors (three windings— main, run, and auxiliary).

The run winding is in the circuit at all times. The start winding is in parallel with the run winding. The start winding is only in the circuit at instant start up. When the motor is at about 60 percent of its rpm, the start winding drops out of the circuit. Exceptions to the above sentence are the PSC motor (permanent split capacitor motor), CSR motor (capacitor start run motor) and the shaded-pole motor (Fig. 5-1).

On full-hermetic and semihermetic compressor motors, switching out of the start winding is done by a relay outside of the motor case. The starting contact points would not hold up inside the motor case because of the oil, freon, moisture and hydrochloric acid.

Suppose that in Fig. 5-2 the resistance of the circuit was 1 ohm and volts were 120. Current will equal what?

$$I = \frac{E}{R} = \frac{120}{1} = 120 \, amps$$

At the instant of start, the current will be very high because the only limitation is dc resistance. If the rotor is locked and you cannot come up to rpm, so as to create back electromotive force (EMF) which bucks applied voltage and limits current, the branch circuit breaker or fuse at the panel will open. Therefore, a motor is an electrical device that has two currents: starting current and running current. A motor needs starting and running current protection. Starting current protection is gained by the circuit breaker or panel fuse, which can be dual element or slow blow. At line voltage, the current of a standard fuse or circuit breaker might be as high as

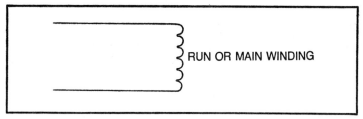

Fig. 5-1. A shaded pole motor.

300 percent of the name plate rating of the compressor motor. This will handle the high starting surge currents, but blow on locked rotor, internal short, or short to ground.

If the motor over-works and the running currents exceed name plate rating plus 33.33 percent while running, the running current protectors to prevent burn out will be the magnetic starter, klixon or internal thermal switch. The wiring of the motor circuits is sized to the name plate running current of the motor from the branch circuit breaker to the motor terminals.

A compressor starting relay is an electrical device to make the starting winding in parallel with the run winding which is across the power lines at no rpm and open the starting winding at full rpm so that the motor will only have the run winding energized by the line voltage at full rpm. I will cover the four starting relays.

CURRENT RELAY

The current relay (Fig. 5-3) uses motor starting current to make a set of contacts. When the motor is up to torque, current falls and the relay—which is sensitive to high starting current—opens contact points. The contact points are normally open when the relay is nonenergized.

POTENTIAL RELAY

The contact points are normally closed on this relay (Fig. 5-4) when it is nonenergized. The relay winding is made parallel with the start winding

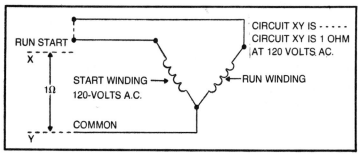

Fig. 5-2. Start and run windings.

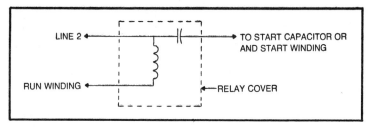

Fig. 5-3. The current relay.

and connected to nonenergized lines 1 and 2. At the instant of start, line voltage is down and the start winding is in the circuit. It is the back EMF potential or back EMF voltage of the start winding that energizes the coil of the potential relay. The contact points open and disconnect the start winding from the power lines. The back EMF voltage of the start winding is normally above the applied line voltage.

At the instant of start, the current is high and the potential (voltage) is low. The contact points of a potential relay (Fig. 5-5) are normally closed. When the compressor is running at full rpm, the start winding generates a potential (could be higher than line potential because you have a magnet turning inside of a field cutting lines of magnetic force). The potential that is generated by the start winding opens the contact points of the potential relay as long as the compressor motor is running. The start winding with the turning rotor is in effect a small alternator. When the contact points of the potential relay open, the start winding and the start capacitor are no longer a circuit from line 1 to line 2.

Figure 5-6 shows a potential relay with two overload klixons at letters A and B. Klixon B has a steel disk that spans two contact points No. 1 and No. 2. Klixon A has a small heater element underneath and in series with the steel disk across contact points No. 1 and No. 3. The heater element makes this klixon a little more current-sensitive.

Letter "O" is an internal thermal switch that is buried in the windings of the compressor motor and provides additional running current protection. Switches A, B, & O provide running current protection. The symbol

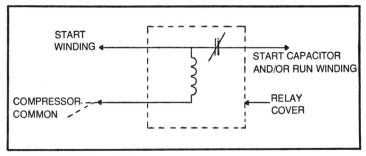

Fig. 5-4. The potential relay.

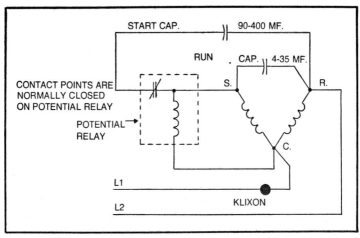

Fig. 5-5. A CSR compressor motor with a potential relay. CSR is a capacitor start run motor.

between numbers 2 and 5 stands for coil. L1 and L2 stands for line 1 and line 2.

There is a little bleed-off (resistor D) that is wired across the start capacitor when the capacitor is not energized. The letters C, S, and R are the motor terminals of the compressor motor.

C is common.

S is start.

R is run.

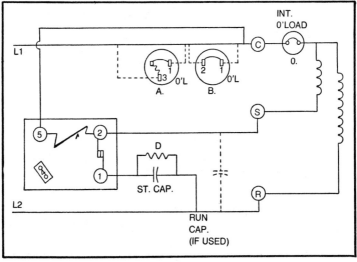

Fig. 5-6. A potential relay with a two-terminal external or internal overload.

In case of overload, switches A, B, & O will take the compressor motor off the line. When the klixons A and B cool down, they automatically reset themselves and the warped steel klixon disk will pop closed across the contact points and reset itself and start the compressor.

When the compressor windings cool down, the internal overload (O) that is buried in the compressor motor windings, will reset itself and start the compressor.

Figure 5-7 is a typical wiring of potential relay with a 3-wire klixon. The steel disk between No. 1 and No. 2 will open if the current of the run winding (Letter R to C) becomes too high. Klixon A will also open if the little heater element, No. 2 to No. 3, gives off too much heat to the steel disc inside the klixon A or because too much current is consumed by the start winding (letter S to C).

The klixon steel disc will become warped and pop open on contacts 1 and 2 because of too much heat. When it cools down after 3 to 5 minutes, it resets itself and you can attempt to start compressor motor again.

Note: This is the only klixon wiring application that does not connect to common. *All* other klixons connect to common C. Figure 5-8 is a 4-wire klixon.

The line wire L1 connects to 4, the current travels through the warped disc to 3 and then you have a connection to common. This klixon has a separate circuit that is not connected electrically to the line circuit. We leave C1 and travel through thermostat B. Thermostat B connects to 1 and then we travel through the separate contact points and leave the klixons at point 2 to C2. The circuit C1 to C2 is a low voltage control or warning circuit and is separate from the line voltage.

In Fig. 5-9, the klixon has been replaced with circuit breaker A rated at 125 percent of full-load current of compressor motor. Starting current protection is gained by panel branch circuit breaker. Circuit breaker A is fused down to give running current protection to the compressor motor. The main disadvantage to this wiring plan is that the circuit breaker does

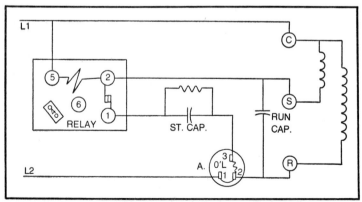

Fig. 5-7. A potential relay with a three-terminal external overload.

40

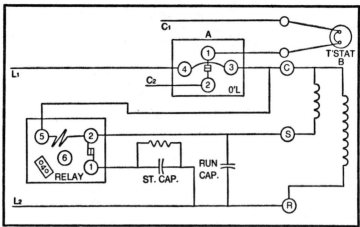

Fig. 5-8. A potential relay with an internal thermostat and external overload.

not reset itself like the klixon. You could lose all your food if breaker **A** failed over a weekend and no one was around to reset breaker **B**. Substitute a 2-wire klixon with the same horsepower rating as the compressor motor in lieu of circuit breaker A. C_1 and C_2 is the temperature-control circuit with overload protection.

HOT-WIRE RELAY

The hot wire relay is an electrical device that has a chrome nickle wire about 4 inches long that gives off sensible heat to two bimetal sensitive contact points. See Fig. 5-10. No. 1. At no voltage, both contact

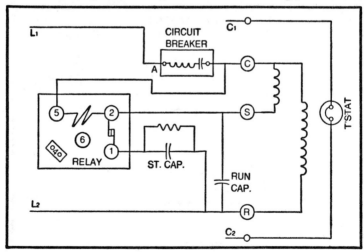

Fig. 5-9. A potential relay with an internal thermostat and an external circuit breaker. C_1 and C_2 is the temperature-control circuit.

points are made. No. 2. When the compressor is up to full rpm, a heat is developed by the chrome nickle wire that warps the bimetal start contact spring and opens the start contact points and holds them open as long as the compressor is running.

The chrome nickle wire is wired in series with the run winding of the compressor motor and the current that is used by the compressor motor travels through the chrome nickle wire and gives off the sensible heat. The compressor motor is limiting the current that is traveling through the wire.

If the compressor motor works too hard, locked rotor occurs or the compressor does not have the pressures equalized on start up. The compressor motor develops too much heat on the hot wire and both the start and the run springs are warped and the start and run contacts open until the hot wire cools down and resets. About 2 to 4 minutes is the reset time.

You will not find a klixon with the hot wire relay because the relay can act as a klixon as well as a starting device. Note: A klixon is a electrical switch that will open on excessive motor current or it will open on excessive temperature of the motor case. Klixons are also used as temperature limiting devices on strip heaters, fan, and high limit switches on furnaces. The klixon has the capability to reset itself automatically, after it cools down. Normally it takes 2 to 4 minutes. The klixon is always wired to common of the compressor motor terminal and the klixon will always point to common. As long as the compressor is running, the hot wire will give off heat and open the start contact points. If you have trouble with the compressor, the hot wire will use more current and give off more heat and warp open the bimetal strip and open the run contact points. The hot wire relay gives running current protection and for this reason you will not find a klixon used with a hot wire relay. Capacitor from 4 to 40 microfarads is a run capacitor. Capacitor from 90 to 400 microfarads is a start capacitor.

PSC means permanent split capacitor compressor motor (Fig. 5-11). This compressor motor has a run capacitor that is always wired between start and run windings. Voltage will travel through the run cap and fire the

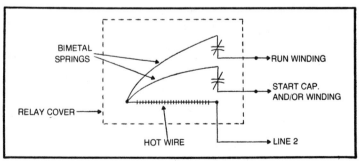

Fig. 5-10. A hot-wire relay.

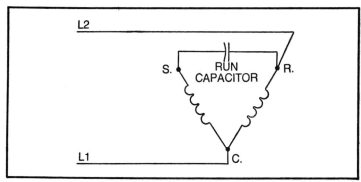

Fig. 5-11. A PSC compressor motor diagram.

start winding first and the run winding second. We have a push push or a sort of 2-phase motor with the PSC motor.

Figure 5-12 shows a standard wiring plan of a capacitor start run compressor motor. We have both windings in parallel at no rpm. We have both capacitors in parallel at no rpm.

The klixon gives running current protection. The temperature sensor switch gives us protection from the windings or compressor dome from getting too hot. A low pressure cut out switch will give us protection if the system gets too cold or the system is low on refrigerant. Refrigerant cools the windings of the compressor motor. A high-pressure, cut-out switch will keep the compressor motor from overworking, too much current will burn out the motor.

SOLID-STATE RELAY

The solid-state relay (Fig. 5-13) is an electrical device that will switch out the start winding by high resistance caused by heat that is generated by the starting current plus heat that is generated by a small trickle current while the motor is running. The solid-state relay must not be too warm on motor start up. Otherwise, the warm temperature will cause too high a resistance and the start winding will not be energized. You will burn up the run winding.

With a solid-state relay, it is imperative that a klixon be inserted in series with the line that is connected to common of the compressor motor terminal. Example: The compressor motor is running. I unplug the compressor motor and allow it to stop. There is no rpm. Next I plug into power. The solid-state is too warm. It will take the solid-state relay 3 minutes to cool down to drop resistance. The klixon will go out because there is too much current on the run winding. Therefore, the klixon will go out, open the line voltage to the compressor for about 2 to 4 minutes, allow the solid-state relay to cool down, resistance drops on the solid-state relay, the klixon resets automatically and if the plug is made the compressor will start.

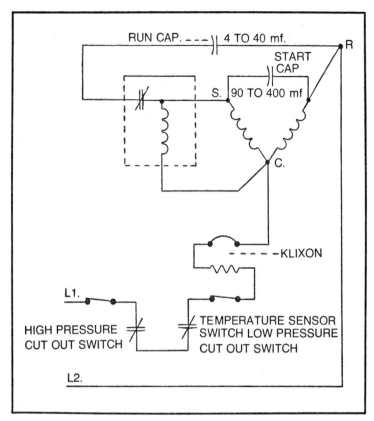

Fig. 5-12. A potential relay where the relay contacts are normally closed. The relay coil is in parallel with the start winding of the compressor motor.

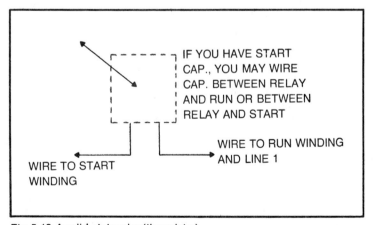

Fig. 5-13. A solid-state relay (thermister).

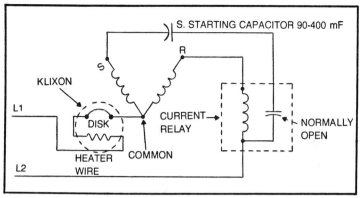

Fig. 5-14. The klixon has a warped steel disk with a chrome nickel heater wire.

Rate the klixon current range at 125 percent to 133 percent of the name plate or running current of the compressor motor. The klixon is a poor man's magnetic starter and gives running current protection while the circuit breaker or panel fuse gives starting current protection. Wiring to the compressor is sized to compressor motor name plate current. For wiring, consult Tables 1-3, 1-4 and 1-5. Undersized wiring can drop voltage too much, burn out compressor motors, and cause fires. In addition, it is against all building codes.

As in Fig. 5-14, the high starting current of the compressor energizes the current relay coil and closes the contact points of the current relay. After the compressor is running, back EMF is developed, starting currents drop to running currents, the current relay coil does not have starting current magnetic flux, the contact points open, and the start winding with the start capacitor is out of the circuit while the compressor

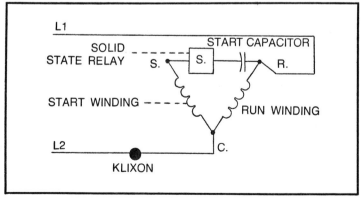

Fig. 5-15. This is standard wiring for a capacitor start compressor with a solid-state relay.

is running. Note: The current relay coil is in series with the run winding of the compressor motor. The current relay contact points are normally open. All of the relays have a klixon except the hot wire relay. The klixon is always wired to common. The klixon points to common.

When the relay is cold the resistance is low. When you start the compressor, the relay becomes warm and develops a very high resistance which stops the flow of current to the start winding. As long as the compressor is running, you will have a trickle current through the solid-state relay which will keep the relay warm with high resistance which stops the flow of current to the start winding. The relay is nonpolarized and can be hooked up to the start capacitor and start winding either way.

The relay and start capacitor can be turned around and the compressor will still work. *Most important:* The solid-state relay (Fig. 5-15) can be used in place of all the other hundreds of relays. However, do not exceed the horsepower rating of the solid-state relay. Always use a klixon for running current protection because the solid-state relay does not have running current protection.

Chapter 6
Compressor Motor
Characteristics

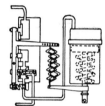

Motor starting currents will normally be between 300 percent to 500 percent of the full-load currents of the motor. Wiring of the compressor is sized to the full-load currents of the motor. There can be more than one motor on an electrical circuit. In that case, you will add the running currents of the motors in the circuit and select the size wiring that is above the total running currents of all the motors in the circuit. Place the wiring inside of a raceway that will give protection and provide a ground. Exceptions are romex and cable (cords). Romex and cable have no raceway. A raceway holds the wire and can be flex, conduit (rigid), wiremold, plugmold, etc. Consult the wiring and conduit tables of this book.

CIRCUIT BREAKER

The circuit breaker or fuse can be sized at 300 percent of the full-load currents of the motors in the circuit. Many times, the circuit breaker or time delay fuse will be sized at less than 300 percent because of the time delay function. The motors have reached rated rpm and the current has dropped before the time delay has expired. Another way to say this is that you have gone from starting current to running current before time delay has expired.

The klixon, or magnetic starter heaters, are sized at 110 to 133⅓ percent of the full-load currents of the motors. The circuit breaker at the panel gives starting current protection. An example is that the circuit breaker or fuse will blow and disconnect the motors from the branch circuit line if there is a locked rotor. A locked rotor can be checked against Table 3-3. The klixon (poor man's magnetic starter) or magnetic starter will give

47

running current protection. An example is condenser fan motor failure. The head pressures become too high and this overworks the compressor motor. The magnetic starter or klixon will take the compressor out of the circuit. Figure 6-1 shows the wiring of an average motor in an electrical circuit.

COILS

Three-phase motors (Figs. 6-2 through 6-5) have six coils of which normally nine leads are carried outside the motor case and terminate inside the motor make-up box that is mounted outside the motor case. These nine leads are numbered from 1 to 9. The electrician will run his conduit to the make-up box, pull his wires from the disconnect switch or magnetic starter to the motor make-up box and wire up the motor by the motor name plate guide.

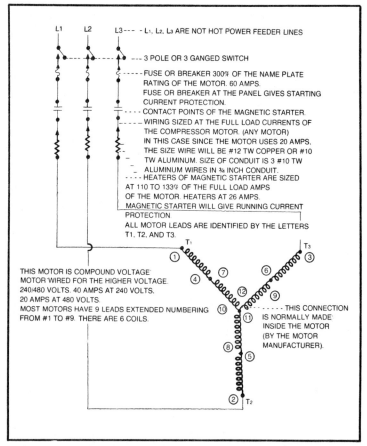

Fig. 6-1. Wiring for an average motor circuit.

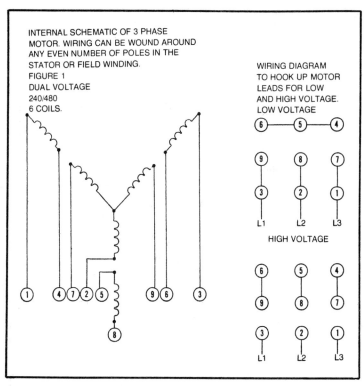

Fig. 6-2. Wiring diagrams.

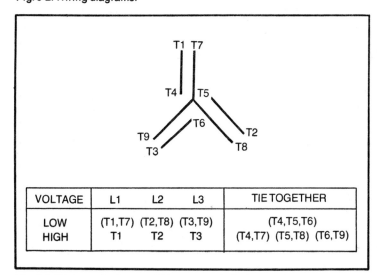

VOLTAGE	L1	L2	L3	TIE TOGETHER
LOW	(T1,T7)	(T2,T8)	(T3,T9)	(T4,T5,T6)
HIGH	T1	T2	T3	(T4,T7) (T5,T8) (T6,T9)

Fig. 6-3. Wiring a single-speed, Y-connected, dual-voltage motor.

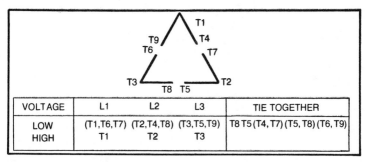

VOLTAGE	L1	L2	L3	TIE TOGETHER
LOW	(T1,T6,T7)	(T2,T4,T8)	(T3,T5,T9)	T8 T5 (T4, T7) (T5, T8) (T6, T9)
HIGH	T1	T2	T3	

Fig. 6-4. Wiring a single-speed, delta-connected, dual voltage.

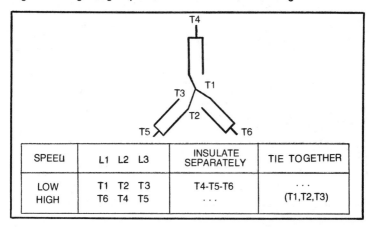

SPEEU	L1 L2 L3	INSULATE SEPARATELY	TIE TOGETHER
LOW	T1 T2 T3	T4-T5-T6	. . .
HIGH	T6 T4 T5	. . .	(T1,T2,T3)

Fig. 6-5. Wiring two-speed, single-winging, variable torque motors.

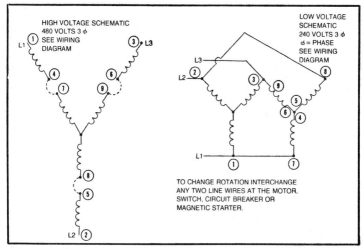

Fig. 6-6. Voltage schematics.

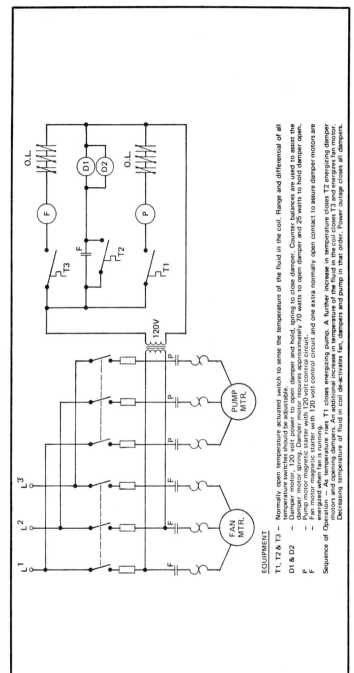

EQUIPMENT

T1, T2 & T3 — Normally open temperature actuated switch to sense the temperature of the fluid in the coil. Range and differential of all temperature switches should be adjustable.

D1 & D2 — Damper motor, 120 volt power to open damper and hold, spring to close damper. Counter balances are used to assist the damper motor spring. Damper motor requires approximately 70 watts to open damper and 25 watts to hold damper open.

P — Pump motor magnetic starter with 120 volt control circuit.

F — Fan motor magnetic starter with 120 volt control circuit and one extra normally open contact to assure damper motors are energized when fan is running.

Sequence of Operation — As temperature rises T1 closes energizing pump. A further increase in temperature closes T2 energizing damper motors and opening dampers. An additional increase in temperature of the fluid in the coil closes T3 and energizes fan motor. Decreasing temperature of fluid in coil de-activates fan, dampers and pump in that order. Power outage closes all dampers.

Fig. 6-7. A closed system water tower with dampers.

51

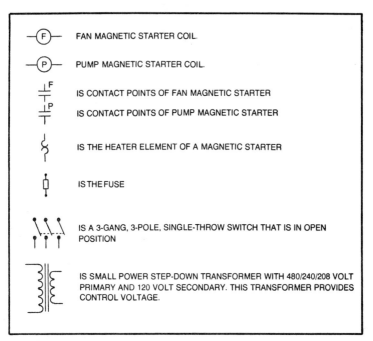

Fig. 6-8. Schematic symbols.

With the wiring, you can select the proper voltage and correct rotation. You can find this wiring in use on semihermetic compressor motors as well as open drive compressor motors. On rotary and centrifical compressors, centrifical water pumps, propeller and squirrel cage fans, rotation is very important. See Fig. 6-6.

The type of wiring and control shown in Fig. 6-7 will give excellent control over condenser water temperature. By stabilizing condenser water temperature you can control the head pressure of the air conditioning or refrigeration system.

The pump motor in Fig. 6-6 is the pump motor that takes the water from the tower pan and delivers it to the spray heads and then the water falls over the tube coil inside the tower and drops into the pan. This process continues as long as the pump is on the line.

O.L. stands for overload relay with the three contact points closed. See Fig. 6-8 for additional symbols.

Chapter 7
Motor Rpm

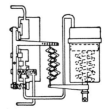

With a two-pole motor, each one-half cycle rotor will move to a new set of poles. The rotor has fixed polarity. The field will change polarity on each one-half cycle. New poles are formed. Therefore, on the first one-half cycle the motor will turn one-half rpm to new poles. On the next one-half cycle, it will turn one-half rpm to a new set of poles. One cycle is equal to 1 rpm and 60 cycles per second will equal 60 rpm per second or 3600 rpm. Figure 7-1 represents the cycle sine wave.

POLES

On the first one-half cycle, pole A is plus and pole B is negative. On the second one-half cycle, pole A is negative and pole B is plus. The rotor is a magnet that turns in the center of the field poles A and B. The rotor magnet will turn because unlike poles attract and like poles repel. At first instant of start, pole A is south and pole B is north. The rotor magnet must move because like poles repel. After the rotor has moved away, the rotor north pole is attracted by field pole A which is south. Unlike poles attract. As soon as the rotor aligns itself with field poles A and B, the next one-half cycle, the field poles change polarity and there is at first the repulsion of poles, the rotor magnet moves, and then there is the attraction of poles.

An electric motor is an electrical device whereby there is a rotor (ac voltage) turning inside of a revolving field and in step with the frequency (cycles per second). See Table 7-1. The speed can be changed electrically by:

■ The manufacturer wiring in more poles in the stator (field) winding by switching.

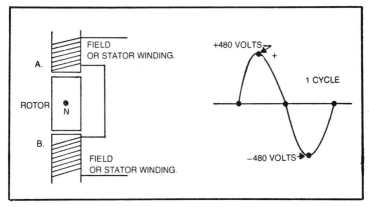

Fig. 7-1. The cycle sine wave.

■ Silicon controlled rectifiers which will chop the voltage wave form, lower motor speed, lower efficiency of the motor, and control current which would rise with the reduced back electromotive force.

When a motor is running in an ac voltage circuit, the motor develops a voltage which bucks the applied line voltage. We call this voltage that is created by the motor back electromotive force or (back EMF). The back electromotive force voltage against the applied line voltage limits the current of the motor.

When you slow a motor down with too much load (work), you reduce back EMF. If current rises above 133 percent of the full-load name plate rating, you will burn out the motor. Other ways to reduce motor speed are gear reducing and belt and pulleys.

All air conditioning requires continuous duty motors (4 plus hours running time service at one time interval). Intermittent duty motors are one-half hour running time or less of service at one time interval. Continuous duty motors could run 24 hours per day all year long if needed.

BELT TENSIONS

You should be able to push a belt in one-fourth inch and push the belt out one-fourth inch. Try not to have too much slap in the belt. If in doubt use a clamp-on ammeter. If you are name plate motor amps to 125 percent name plate motor amps you are OK.

2-pole motor is 3600 rpm with no load.
4-pole motor is 1800 rpm with no load.
6-pole motor is 1200 rpm with no load.
8-pole motor is 900 rpm with no load.
10-pole motor is 720 rpm with no load.
12-pole motor is 600 rpm with no load.

Table 7-1. Motors and rpm.

Chapter 8
Basic Motor Control

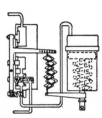

Figure 8-1 is a block diagram of a basic 3-phase, 240/480 volts motor control.

   1. **Circuit Breaker or Branch Circuit Fuse.** Rated at 300 percent of full load current of the motor—name plate rating. The circuit breaker or fuse will give starting current protection. Example is a short in the motor or locked rotor.

   2. **Motor Disconnect Switch.** When the motor is over 50 feet from No. 1 circuit breaker or when the motor is out of sight from No. 1 circuit breaker, a non-fused disconnect switch is required.

   3. **Magnetic Starter.** This electrical device gives running current protection. The magnetic starter heaters are sized at 125-133 percent of the name plate rating motor amperes.

   4. **Compressor Motor.** Full hermetic, semihermetic, or open drive.

   5. **Low Voltage Control Transformer.** The secondary or control voltage is 24 or 120 volts ac. The 24-volt thermostat letter code (in general use):

   R stands for low voltage hot.
   C_1 stands for common return.
   Y stands for 1st stage cool.
   Y^2 stands for 2nd stage cool.
   W^1 stands for 1st stage heat.
   W^2 stands for 2nd stage heat.
   G stands for fan.
Fedders/Westinghouse:
   V stands for low voltage hot.
   F stands for fan.
   C stands for cool.

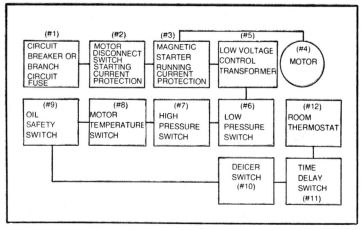

Fig. 8-1. Basic motor control (three-phase, 240/480 volts).

H stands for heat.

L1, L2, L3 stand for line 1, line 2, and line 3.

T_1, T_2, T_3 stand for motor legs 1, 2, and 3.

HEATING & COOLING THERMOSTAT

Figure 8-2 shows a typical heating and cooling thermostat with code letters.

YI is first stage cooling. This is the lead compressor. Most large tonnage ac installations have more than one compressor.

Y2 is second stage cooling. This would be the second compressor that would come on the line automatically.

Letter "O" stands for a cooling damper motor or change-over valve.

RC is a separate low voltage power hot coming from secondary as low voltage transformer 1.

W1 is first stage heating. This could bring on a heat pump or boiler.

W2 is second stage heating. This terminal could bring on a second boiler on the line or a strip heater as the system is not keeping up with the heating requirements.

Letter "B" will feed a heating damper or change-over valve.

6. **Low Pressure Switch.** A safety control to stop the unit if there is a refrigerant leak or too low evaporator temperature resulting in liquid flood back. An obstruction in the refrigeration system will trip the low pressure switch and the high pressure be too high. As a result, the high side switch may trip.

7. **High Pressure Switch.** A safety control to stop the compressor condenser section if there is head pressure in excess of 375-400 psi. High head pressures are caused by a failure of the outside condenser fan, plugged air filters, obstruction in the refrigerant line, or over charge of refrigerant.

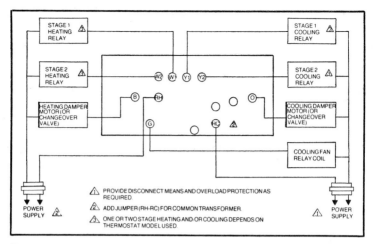

Fig. 8-2. A typical heating and cooling thermostat with code letters.

8. **Motor Temperature Switch.** A safety control to stop the unit when the compressor motor windings or case temperature become too hot. You should be able to touch the motor of the compressor and be able to count to 10 without burning your hand. If you cannot do this test, you have something wrong. The switch can be inside the motor with pins or terminals extended or an outside klixon switch.

9. **Oil Safety Switch.** This switch will stop the unit if there is not enough oil to lubricate. The oil leaves the crank case of the compressor to the evaporator or condenser depending upon pressure of the refrigerant. Check the switch by ohmmeter and the oil level by the sight glass on the crank case of the compressor.

10. **Deicer Switch.** This switch will stop the unit should it drop below 34°F. on the chill water temperature. Low chill water temperature is caused by chill water pump failure or system low on refrigerant.

11. **Time Delay Switch.** This switch will prevent short cycling of the compressor unit. It is set normally for a delay of 2-3 minutes duration.

12. **Thermostat.** A switch to control temperature by cycling compressor condenser unit. The temperature can be controlled by temperature, gas pressure, or air pressure.

For a gas pressure switch, cut out is desired temperature less 10 degrees F. Differential is 10 psi. Cut in is cut out plus 10 psi. For close temperature tolerance, make the differential 5 psi. Be careful not to short cycle compressor condenser section (unit of for less than 3 minutes).

A WIRING SCHEMATIC & A LADDER SCHEMATIC

Figure 8-3 is a wiring schematic. It shows the actual layout of the components. A ladder schematic is drawn like a ladder when the loads are in parallel (Fig. 8-4).

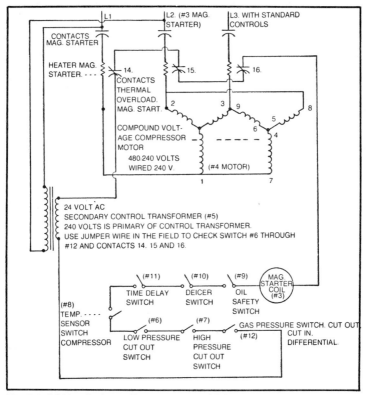

Fig. 8-3. A 240-volt, three-phase compound voltage compressor motor.

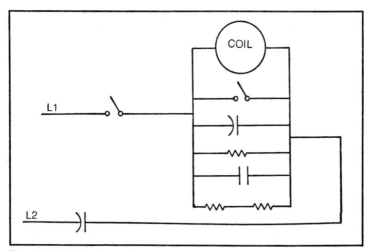

Fig. 8-4. A ladder schematic.

Chapter 9
The Basic Mechanical
Refrigeration System

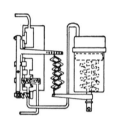

The basic mechanical refrigeration system has four main components. There can be accessories in addition to the four main components. An accessory is a part that is not needed but used to refine the work being done.

Full hermetic means a sealed motor compressor unit. Semihermetic means that you have a nut and bolt constructed motor compressor unit which can be taken apart and overhauled. A compressor with an outside drive means that there is an extended shaft with a seal and this shaft is driven by an outside power source. This outside compressor shaft can be directly coupled to a motor or belt and pulley method. See Fig. 9-2.

The purpose of the compressor is to keep the refrigerant moving and to raise the pressure of the refrigerant above the condensing pressure. The condenser is a heat exchanger that the refrigerant enters as a high-pressure, superheated vapor and leaves as a high-pressure liquid. When the heat is removed from the refrigerant in the condenser by the air or water cooling it, the refrigerant changes from a vapor at the top to liquid at the bottom.

The heat that is removed from the condenser is the heat of compression and the heat of evaporation. We need the condenser to manufacture liquid refrigerant. The expansion devices will only drop pressure on liquid refrigerants. We can not feed an expansion device with vapor and expect a pressure drop.

The expansion device is a part in a refrigerant system to change a high-pressure liquid to a low-pressure, particle-sized vapor. We need to drop pressure in order to drop temperature. An exception is a high-side float. Some expansion devices are the capillary tube, thermostatic expansion valve, automatic valve, and float. The expansion device feeds

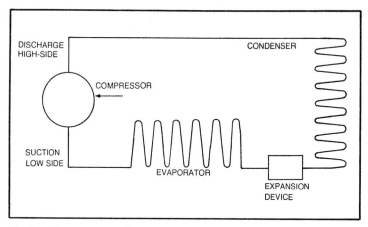

Fig. 9-1. The compressor is a vapor pump.

low-pressure refrigerant to the evaporator. The evaporator is the cooling coil. An exception is the heat pump while on heating cycle.

The evaporator is a heat exchanger that will absorb heat into the refrigerant that is circulating through it. An exception is the heat pump.

TYPES OF HEAT EXCHANGERS

Straight or coiled tube.
Tube on plate.
Tube and fin.
Tube and wire.
Tube on a tube.
Tube in a tube.
Shell and tube.
Shell jacket such as environmental chambers.
Sun on glass plates plus water (solar).
Sun on gravel under glass (solar).
Large wheel of media turning in supply and return air.
Tubes with salt water solutions.
Tubes with special chemicals.
Tubes with gels.

The last four heat exchangers could be known as recuperators which are mechanical heat recovery devices.

Heat can be transferred by conduction, convection, or radiation. *Conduction* is where the heat exchanger touches or comes into contact with what is to be cooled. *Convection* is natural air currents. Hot air rises and cool air falls. *Radiation* is heat that is boiled off or thrown off a surface of a heat exchanger. An example would be a strip heater or pot bellied stove. Other examples would the sun and the heat in an inclosed car on a sunny day. Heat travels from hot to cold.

REFRIGERANT

Figure 9-2 shows the basic mechanical refrigeration system. We leave the compressor on the high side with a high-pressure, super-heated vapor. Heat of compression plus the heat of evaporation. At the beginning of the condenser, the refrigerant is super-heated, high-pressure vapor. The bottom of the condenser we leave with a high-pressure liquid. The line side of the expansion device is fed with a high-pressure liquid. We leave the expansion device (cap, tube, valve, or float) with low-pressure, particle-sized vapor. The evaporator is fed with low-pressure particle-sized vapor and the refrigerant leaving the evaporator is low-pressure, super-heated vapor. The refrigerant becomes heated by the heat load on the machine. The four processes involved are compression (compressor), condensation (condenser), expansion (expansion device), and evaporation (evaporator).

ACCEPTABLE MACHINE TEMPERATURE/PRESSURES

In typical building air conditioning, all evaporators have a temperature of the refrigerant inside of 40 degrees Fahrenheit. Therefore, the refrigerant suction or low-side pressure for general air conditioning is:

69 psi for R 22 refrigerant—high side 225 psi.

37 psi for R 12 refrigerant—high side 125 psi.

59 psi for R 717 refrigerant (ammonia)—high side 400 psi and up.

80 psi for R 502—16 inches of mercury for R 11 refrigerant with a high side pressure not to exceed 15 psi. Normal high side pressure for a R 11 system is about 5 psi.

With automotive air conditioning, the low side will be 10 to 30 psi and the high side will be between 125 to 300 psi (factory air). If you have an add-on unit, try to make suction 37 psi and high side 125 to 300 psi. Make sure that the sight glass is full.

For freezers, the suction or low side will be 1 to 2 psi for R 12 with high side of 125 psi and 10 to 13 psi for R 22 with high side of 225 psi. Dual

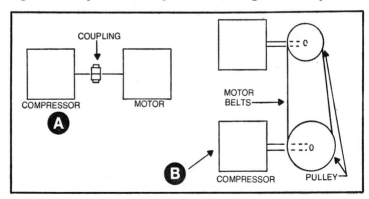

Fig. 9-2. A basic refrigeration system.

temperature refrigerators have a low side of 6 to 8 psi for R12 with a high side pressure of 125 psi. 0 to −5 F freezer compartment, 40 to 45 F cooler section.

COOLERS

For coolers holding milk, beer or similar liquids, the pressures are:
Low side is 37 psi for R 12—high side is 125 psi.
Low side psi is 69 psi for R 22—high side is 225 psi.

The temperature of a cooler is between 40 to 45 degrees Fahrenheit. Commercial and industrial refrigeration temperature within refrigerated space is controlled by gas pressure switches. For example, general air conditioning skin temperature of evaporators is 50 degrees Fahrenheit. In order to have this skin temperature we must carry the refrigerant inside the evaporator 10 degrees Fahrenheit colder than the coil. This is 40 degrees Fahrenheit. The corresponding temperature for 40 degrees is 69 psi if the refrigerant is R 22. Make the differential 10 psi on the pressure switch and the cut in at 80 psi.

The differential is the difference between cut out and cut in on the gas pressure operated switch. The skin temperatures of the evaporator coil are controlled by a gas pressure operated switch that cycles the compressor condenser section rather than a temperature sensitive switch.

Cut in is the pressure that we use to start the compressor condenser section. Cut out is the pressure that we use to stop the compressor condenser unit. *Differential* is the time lag or difference between cut out and cut in. For example, we have an evaporator sized and placed in a walk in freezer room. We want to make the room −10 degrees F. The refrigerant in the evaporator will have to be 10 degrees colder than the room temperature. So we set cut out on the switch at .6 psi which is −20 degrees F for R 12 refrigerant.

Cut in will be at 10 psi. Differential will be 10 psi. We pull the room temperature to a little colder than −10 degrees F. The compressor condenser unit cuts out and when the coil temperature rises to 10 psi we pull down the unit again.

If the room is too warm, we will narrow the differential to 5 psi. We must be very careful with the compressor condenser section as to not short cycle it. If the compressor is off for less than 4 minutes, you are short cycling and the differential is too small. Short cycling will shorten the life and eventually burn out the compressor. When the compressor cycles out of the circuit, it must be out for more than 4 minutes. About 6 to 10 minutes would be a minimum time interval. Make the time out as long as possible without dropping the indoor temperature. A long time out allows for the pressures to equalize a little bit before we start up again.

Domestic and residential systems control temperature of coils and rooms by a temperature switch. You want the room at 60 degrees F so you set the thermostat at 60 degrees and that is it. On refrigerators, you pick the proper setting by temperature and that is the temperature.

PRESSURE SWITCHES

Try to carry the refrigerant temperature inside the coil 10 degrees F colder than you want to make the room or case temperature. Make the differential 10 psi. If the temperature rises, lower the differential to 5 psi. If we short cycle, lower refrigerant temperatures in the evaporator are 15 degrees colder than the case with a differential of 10 psi. If it is still warm, go to a differential of 5 psi. Watch out for short cycling (compressor is off for less than 4 minutes) I have run ice cream cases to -30 degrees F and never used the cut out switch. The compressor was running 24 hours a day for 3 weeks as I was unable to purchase a starting relay. The ideal situation is to match the machine tonnage to the heat load and outside ambient air with as long a compressor running time as possible.

CONDENSING PRESSURES

A rule of thumb calls for a pressure of 25 degrees Fahrenheit above outside ambient air for air-cooled condensers. There will be a pressure of 20 degrees Fahrenheit above the outside ambient air for water-cooled condensers. An example is a R 22 system with a water cooled condenser. Where the outside temperature is 100 degrees Fahrenheit, take the outside temperature and add 20 degrees. This will give 260 psi and this would be acceptable. Every day head pressures will vary.

For air-cooled condensers, if you can't remember the rule of thumb, just remember 225 psi is a good average for R22 and 125 psi is a good average R12. There is one exception with R12 and that is with automotive air conditioning. On this, the pressures will be between 125 to 300 psi. Keep in mind that 165 is a good average pressure for air-cooled condensers for R500 and that 250 is a good average pressure for air-cooled condensers for R502.

Water-cooled condensers are very popular on high-tonnage machines because condensing pressures are much lower. For example, on an air-cooled condenser R22 system operating at 255 psi, the same system on a water-cooled condenser would have a pressure of 190 to 200 psi. Most refrigeration from 75 to 1000 tons will probably have water cooled condensers. Air cooled condensers will loose efficiency when the outside ambient air goes over 100 degrees F.

In very hot climates, it is possible to find precoolers on air-cooled condensers. A precooler could be a swamp cooler placed ahead of the air-cooled condenser. Air-cooled condensers are as large as 1000 tons.

RULES OF THUMB

One ton will handle 400 square feet of room cooling with the ceiling 8 to 10 feet high.

One ton of air conditioning will give you a split between outside air and inside air of 25 degrees Fahrenheit if the square feet is 400 and the ceiling is 8 to 10 feet high. This is with an average building and average insulation.

One horsepower is equal to 1 ton of air conditioning. Many manufacturers will use a slight more than 1 horsepower per ton. One horsepower is equal to 800 watts and 12,000 Btu per hour is equal to 1 ton

Table 9-1. Temperature/Vapor Pressure.

TEMP F	C	R11	R12	R22	R500	R502	R717
−30	−34.4	27.8	5.5	4.9	1.2	9.4	
−28	−33	27.7	4.3	5.9	.1	10.5	.0
−26	−32	27.5	3.0	6.9	.9	11.7	.8
−24	−31	27.4	1.6	7.9	1.6	13.0	1.7
−22	−30	27.2	.3	9.0	2.4	14.2	2.6
−20	−29	27.0	.6	10.2	3.2	15.5	3.6
−18	−28	26.8	1.3	11.3	4.1	16.9	4.6
−16	−27	26.6	2.1	12.5	5.0	18.3	5.6
−14	−26	26.4	2.8	13.8	5.9	19.7	6.7
−12	−24	26.2	3.7	15.1	6.8	21.2	7.9
−10	−23	26.0	4.5	16.5	7.8	22.8·	9.0
−8	−22	25.8	5.4	17.9	8.8	24.4	10.3
−6	−21	25.5	6.3	19.3	9.9	26.0	11.6
−4	−20	25.3	7.2	20.8	11.0	27.7	12.9
−2	−19	25.0	8.2	24.4	12.1	29.4	14.3
0	−18	24.7	9.2	24.0	13.3	31.2	15.7
2	−17	24.4	10.2	25.6	14.5	33.1	17.2
4	−16	24.1	11.2	27.3	15.7	35.0	18.8
6	−14	23.8	12.3	29.1	17.0	37.0	20.4
8	−13	23.4	13.5	30.9	18.4	39.0	22.1
10	−12	23.1	14.6	32.8	19.7	41.1	23.8
12	−11	22.7	15.8	34.7	21.2	43.2	25.6
14	−10	22.3	17.1	36.7	22.6	45.5	27.5
16	−9	21.9	18.4	38.7	24.2	47.7	29.4
18	−8	21.5	19.7	40.9	25.7	50.1	31.4
20	−7	21.1	21.0	43.0	27.3	52.5	33.5
22	−6	20.6	22.4	45.3	28.9	54.9	35.7
24	−4	20.1	23.9	47.6	30.6	57.4	37.9
26	−3	19.7	25.4	50.0	32.4	60.0	40.2
28	−2	19.1	26.9	52.4	34.2	62.7	42.6
30	−1	18.6	28.5	54.9	36.0	65.4	45.0
32	0	18.1	30.1	57.5	37.9	68.2	47.6
34	1	17.5	31.7	60.1	39.9	71.1	50.2
36	2	16.9	33.4	62.8	41.9	74.1	52.9
38	3	16.3	35.2	65.6	43.9	77.1	55.7
40	4	15.6	37.0	68.5	46.1	80.2	58.6
42	5.6	15.0	38.8	71.5	48.2	83.4	61.6
44	6.7	14.3	40.7	74.5	50.5	86.6	64.7
46	7.8	13.6	42.7	77.6	52.8	90.0	67.9
48	8.9	12.8	44.7	80.8	55.1	93.4	71.1
50	10	12.0	46.7	84.0	57.6	96.9	74.5
52	11	11.2	48.8	87.4	60.1	101	78.0
54	12	10.4	51.0	90.8	62.6	104	81.5
56	13	9.6	53.2	94.3	65.2	108	85.2
58	14	8.7	55.4	97.9	67.9	112	89.0
60	16	7.8	57.7	102	70.6	116	92.9
62	17	6.8	60.1	105	73.5	120	96.6
64	18	5.9	62.5	109	76.3	124	101
66	19	4.9	65.0	113	79.3	128	105
68	20	3.8	67.6	117	82.3	132	110

Table 9-1. Temperature/Vapor Pressure (continued from Page 64).

F	C	R11	R12	R22	R500	R502	R717
70	21	2.8	70.2	121	85.4	137	114
72	22	1.6	72.9	126	88.6	141	119
74	23	.5	75.6	130	91.8	146	123
76	24	.3	78.4	135	95.1	150	128
78	26	.9	81.3	139	98.5	155	133
80	27	1.5	84.2	144	102	160	138
82	28	2.2	87.2	148	106	165	144
84	29	2.8	90.2	153	109	170	149
86	30	3.5	93.3	158	113	175	155
88	31	4.2	96.5	163	117	180	160
90	32	4.9	99.8	168	121	186	166
92	33	5.6	103	174	125	191	172
94	34	6.3	107	179	129	197	178
96	35	7.1	110	185	133	203	184
98	36	7.9	114	190	137	208	191
100	37	8.8	117	196	141	214	197
102	39	9.6	121	202	146	220	204
104	40	10.5	125	208	150	227	211
106	41	11.3	129	214	155	233	217
108	42	12.3	132	220	159	239	225
110	43	13.2	136	226	164	246	232
112	44	14.2	141	233	169	252	240
114	45	15.1	145	239	174	259	248
116	46	16.1	149	246	179	266	255
118	47	17.2	153	253	184	273	264
120	49	18.2	158	260	189	280	272
122	50	19.3	162	267	195	288	280
124	51	20.5	167	274	200	295	289
126	52	21.6	171	282	206	303	
128	53	22.8	176	289	212	310	
130	54	24.0	181	297	217	318	

of air conditioning while 1 pound of refrigerant should handle 1 ton of refrigeration exclusive of liquid receivers, accumulators, driers and other accessories. Machines that use too much energy, have a poor design, or are of poor quality will vary from the above relationships.

The abbreviation for British thermal unit is Btu. It will take 1 Btu to raise the temperature of 1 pound of water 1 degree Fahrenheit. Specific Heat is the Btu needed to raise the temperature of a substance 1 pound by weight 1 degree Fahrenheit.

Every substance will have a different specific heat. Heat transfer is greatest when the temperature between heat exchanger and product load is greatest. An example is the heat exchanger of a home furnace which is around 900 degrees Fahrenheit. Air entering equals 50 degrees Fahrenheit and the same air leaving will equal 100 to 120 degrees Fahrenheit at 800 to 1100 cubic feet per minute (CFM).

FAHRENHEIT TO CENTIGRADE

The temperature shown in Table 9-1 (Fahrenheit to Centigrade) are within 1 degree of true value. This is good enough for field applications. The conversion from Fahrenheit to Centigrade is not a straight-line function. For exact values, see Tables 9-2A and 9-2B. I have selected the range of −30 degrees to 130 degrees Fahrenheit for vapor pressures because this is the most practical for field use. Vapor pressures are given in pounds per square inch.

Using Fig. 9-3(A), you can do a conversion of Fahrenheit or Centigrade to 3600 degrees C. You can use Fig. 9-3(B) to do a conversion of Fahrenheit to Centigrade to −459.72 degrees F or −273.18 degrees C.

Table 9-2A. Temperature Conversion.

Fahrenheit to Centigrade			
−30	−34.44	50	10
−28	−33.33	52	11.11
−26	−32.22	54	12.22
−24	−31.11	56	13.33
−22	−30	58	14.44
−20	−28.89	60	15.56
−18	−27.78	62	16.67
−16	−26.67	64	17.78
−14	−25.56	66	18.89
−12	−24.44	68	20
−10	−23.33	70	21.11
−8	−22.22	72	22.22
−6	−21.11	74	23.33
−4	−20	76	24.44
−2	−18.89	78	25.56
0	−17.78	80	26.67
2	−16.67	82	27.78
4	−15.56	84	28.89
6	−14.44	86	30
8	−13.33	88	31.11
10	−12.22	90	32.22
12	−11.11	92	33.33
14	−10	94	34.44
16	−8.89	96	35.56
18	−7.78	98	36.67
20	−6.67	100	37.78
22	−5.56	102	38.89
24	−4.44	104	40
26	−3.33	106	41.11
28	−2.22	108	42.22
30	−1.11	110	43.33
32	0	112	44.44
34	1.11	114	45.56
36	2.22	116	46.67
38	3.33	118	47.78
40	4.44	120	48.89
42	5.56	122	50
44	6.67	124	51.11
46	7.78	126	52.22
48	8.89	128	53.33
		130	54.44

Table 9-2B. Centigrade to Fahrenheit.

°C	°F
−34	−29.2
−32	−25.6
−30	−22
−28	−18.4
−26	−14.8
−24	−11.2
−22	−7.6
−20	−4
−18	−0.4
−16	3.2
−14	6.8
−12	10.4
−10	14
−8	17.6
−6	21.2
−4	24.8
−2	28.4
0	32
2	35.6
4	39.2
6	42.8
8	46.4
10	50
12	53.6
14	57.2
16	60.8
18	64.4
20	68
22	71.6
24	75.2
26	78.8
28	82.4
30	86
32	89.6
34	93.2
36	96.8
38	100.4
40	104
42	107.6
44	111.2
46	114.8
48	118.4
50	122
52	125.6
54	129.2
56	132.8

Steel melts at 5000 degrees F. The temperature of a welding torch is 6000 degrees F. Easy-flow 45 silver solder melts at 1400 degrees F. Still floss silver solder melts at 1900 degrees F. Lead solder melts at 400 degrees F. Freezing at 32 degrees F equals 0 degrees C. Boiling at 212 degrees F equals 100 degrees C.

Table 9-3. Wind Chill Chart.

Speed in MPH	45	40	35	30	25	20	15	10	5	0	-5	-10	-15	-20
4	45	40	35	30	25	20	15	10	5	0	-5	-10	-15	-20
5	43	37	32	27	22	16	11	6	0	-5	-10	-15	-21	-26
10	34	26	22	16	10	3	-3	-9	-15	-22	-27	-34	-40	-46
15	29	23	16	9	2	-5	-11	-18	-25	-31	-38	-45	-51	-58
20	26	19	12	4	-3	-10	-17	-24	-31	-39	-46	-53	-60	-67
25	23	16	8	1	-7	-15	-22	-29	-36	-44	-51	-59	-66	-74
30	21	13	6	-2	-10	-18	-25	-33	-41	-49	-56	-64	-71	-79
35	20	12	4	-4	-12	-20	-27	-35	-43	-52	-58	-67	-74	-82
40	19	11	3	-5	-13	-21	-29	-37	-45	-53	-60	-69	-76	-84
45	18	10	2	-6	-14	-22	-30	-38	-46	-54	-62	-70	-78	-85
50	17	9	0	-7	-17	-24	-31	-39	-47	-56	-63	-71	-79	-88

Source: U.S. Army   NOTE: Relative wind speeds greater than 50 mph have little further additional effect this is bare skin temperature, lower from the result of water evaporation.

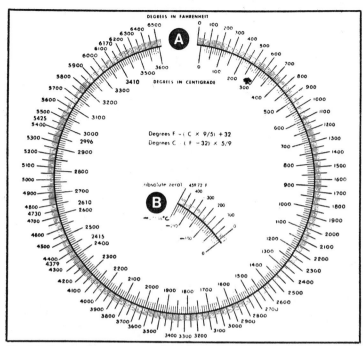

Fig. 9-3. Conversion charts.

WIND CHILL FACTOR

When the humidity is quite high, you will feel colder. If the humidity is low, you will not feel as cold. Wind speed in a cold air duct is approximately 11 miles per hour in an average 2-story or 3-story building. Wind speed in a cold air duct is approximately 20 mph in an average high-rise building.

You will need five times the volume of air for cooling than heating. The reason is that the differential of supply air to room temperature is much greater for heating than cooling. Example: room temperature at 72° F. Supply heating can be 95-120 degrees F. Supply cooling is 20 degrees less than room temperature. Cooling differential is 20 degrees F on the average. Heating differential is 40 degrees to 50 degrees on the average. Wind speed in a hot air deck is 5 mph or less.

Any air stream in a room that is 15 mph or more is considered a draft. Advantage of ducts with the insulation inside the ducts is that air can be moved at higher speeds than 11 mph and there will be no noise. A disadvantage of ducts with the insulation on the inside is that when the felt barrier that is in contact with the air stream (moving through the duct) breaks down, the asbestos that is in the insulation will be picked up by the air stream, become airborne, and enter the area where people are living. When the air is breathed, you would have a health hazard.

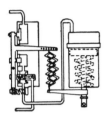

Chapter 10
Checking a
System for Leaks

The first step is to look for oil spots on the refrigerant system. There is 2 percent to 4 percent oil mixed with the refrigerant throughout the entire system and for some reason it will always come out and show where the leak is. If you see no oil, pressurize the system above 30 psi with refrigerant of the system or nitrogen. Take a small bottle and mix a soap solution using one-third soap to two-thirds water. Take a small paint brush and soap up every flare, hose, clamp, silver solder joint, compressor head bolt, gaskets, and sight glass. Check for fatigue pipes, corrosion and copper oxide (copper oxide eats through copper and is green in color). Steel condensers will rust out. Remember that refrigerant does not wear out, but just leaks away. Tighten up flare nuts, bolts and clamps. Blow the charge and silver solder or aluminum solder the crack or pin hole. It is not possible to repair the leak under pressure with solder. You will always have a pin hole. If the heat exchanger has too many holes or the sight glass leaks, you may have to replace the bad parts.

In the case of R11 systems, many times a service mechanic will have success in getting rid of leaks on the low side by painting over the area with varnish while the unit is running. Low side pressure is 16 inches of HG. The negative pressure will not pin hole through the varnish. Other methods to check for leaks is with a halide torch which will glow green in color when you are over the leak with the sensor hose. Some leak checkers will make noise when you find the leak with their sensors or others will have lights that flash when you place the sensor on the leak.

If you have a leak in a closed room, some of these leak checkers are too sensitive and will not be satisfactory. They will sense the refrigerant in the air and show leaks everywhere.

When the leak is repaired, next vacuum the refrigerant system. The 24 hour vacuum is the best method. The second method is called triple evacuation. It is the most common vacuum process. Begin with 1 hour of vacuuming the refrigeration system with the vacuum pump. Then take 5 minutes to add refrigerant vapor to the system. Spend another 1 hour of vacuuming the refrigeration system with the vacuum pump. Take another 5 minutes to add refrigerant vapor to the system. Spend 1 more hour vacuuming the refrigeration system with the vacuum pump.

Next, shut off the gauges and see if the refrigeration unit holds the vacuum. Note: when you are vacuuming the refrigeration unit, pull the vacuum from both the high and low side of the system. The vacuum pump goes to the center hose and the hoses hook to the respective sides of the refrigeration system.

When you add vapor to the system between vacuum time, this process is called a *sweep*. Use vapor R12 to blot up moisture within the system and speed up the process. After each vacuum, turn off both the high-side valves and low-side valves on the gauge manifold. Keep the system sealed. Next, turn off the vacuum pump and disconnect the hose from the pump. The bottle of refrigerant is hooked up to the center hose of the manifold. Next, open up the bottle valve and let vapor enter the hose. On the center hose flare that is on the manifold, open it a crack and let some refrigerant escape out and thus purge the air in the center hose with refrigerant. This process is called purging the hose of air. The time to do this takes about 3 seconds.

When the hose is purged of air, sweep the machine with vapor R12 refrigerant. To get vapor out of a refrigerant bottle, draw out the refrigerant from the bottle while the bottle is in the upright position. To draw out liquid, turn the bottle upside down.

After the sweep, turn off the bottle and the refrigerant gauges. The bottle is removed from the center hose. Open both sides of the refrigerant gauges a little bit and let the refrigerant escape. This must be done slowly so as not to lose the system oil.

When the gauges are less than 5 psi, start the vacuum pump up and hook up the center hose and start your hour vacuum. I suggest an oil change on the vacuum pump after two tripple evacuations or one 24 hour vacuum. Many service men will start the vacuum on the system when they no longer hear a hiss of sweep gas leaving the system.

On final charge, have the gauges both closed and the refrigeration system has a vacuum of 28 to 29 inches of mercury vacuum. Let the machine stay under this vacuum for 20 minutes. If the gauges remain the same, you do not have a leak. Select the refrigerant of the system, hook up the center hose to the refrigerant bottle, purge the hose and open the low side valve. After one-half minute, start up the refrigerant system and charge until the sight glass is clear and has no bubbles or pressure charge using the psi (mentioned earlier in this section).

The third vaccum method is the System purge. If the machine is worked on and open to the outside air less than 3 minutes, take the refrigerant bottle and go in on the low side with vapor and open up the service entrance port on the high side and let the flowing refrigerant purge out moisture in the system for 3 minutes. Close the high-side valve, start up the refrigerant system, and go into final charge. Use either sight glass charging or pressure charging.

REFRIGERANT PRESSURES

Boyle's law says that if the temperature of a gas is constant, the pressure is inversely proportional to its volume. If the pressure of the gas is lowered, the volume of the gas expands.

Figure 10-1(A) shows 2 balloons of 1 pound of gas by weight each. Figure 10-1(B) shows the same 2 balloons. Balloon B is placed in a vacuum jar. The vacuum pump lowers the pressure around the outside surface of balloon B. Balloon B expands. Assume Fig. 10-1(C) is a R 12 refrigeration system. Leave the compressor A high side at 125-psi vapor. Leave the condenser at 125-psi liquid and the capillary tube drops the pressure that feeds the evaporator to 1 psi particle sized vapor.

Why is the machine not starved on the low side? We do not starve on the low side because as we drop pressure across the capillary tube form 125 psi liquid to 1 psi vapor (particle sized vapor), the gas in the cooling coil—called the evaporator—expands like the gas in the balloons and the entire inside wall of the evaporator is wet and cold (−20 degrees Fahrenheit all the way to the suction of the compressor).

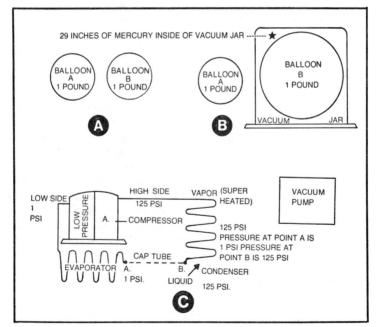

Fig. 10-1. Refrigerant pressures.

Why do we not wind up with all the refrigerant on the high side? We do not wind up with all the refrigerant on the high side because capillary tube machines are critically charged with just enough refrigerant to feed the capillary tube with liquid to drop pressure from 125 psi to 1 psi. All expansion devices such as cap. tubes, expansion valves and high-side floats must be fed with liquid as they will not drop pressure with vapor. Expansion valves are rated in tons with the refrigerant they use stamped on the valve. Cap. tubes come in all different sizes (diameter and lengths). Cap. tubes will handle one-eighth of a ton to 5 tons.

GAY-LUSSAC'S LAW

This law states that if we have a constant volume, temperature is directly proportional to "vapor" pressure. This law explains the saturated vapor temperature/pressure tables. A refrigerant system is a closed system with constant volume. Therefore, I can make the temperature of a coil from -50 F to 200 F by selecting proper pressure and refrigerant vapor from the saturated vapor temperature/pressure tables.

CHARLES' LAW

If we have a constant pressure, the temperature is directly proportional to the volume. This explains why the gas volume is larger on a hot day and less on a cold day.

Figure 10-2(A) shows two balloons of the same weight of gas. For example, let's say that they weigh 1 pound each. In Fig. 10-2(B) we keep balloon No. 1 at the same temperature as before and the gas in balloon No. 2 is heated. This can be done by placing the balloon in the sun. The end result is that balloon No. 2 becomes larger because of Charles' law. This explains why the gas in a charging cylinder has a different volume when the temperature differs. On a cold day, a machine can be over-charged because the volume is more dense by weight than on a warm day.

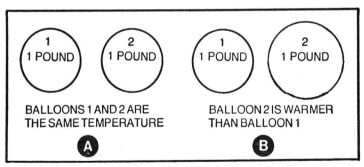

Fig. 10-2. Charles' law.

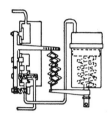

Chapter 11
Problems
With the Enclosed
Refrigerant System

Examples of an obstruction are plugged capillary tubes, bad thermalstatic expansion valves, and plugged condensers. The best type of repairs would be a replacement of the plugged component.

LEAKS

Refrigerant doesn't wear out; it just leaks away. Locate the leak with soap bubbles or a leak detector, repair the leak, vacuum the system and make the final charge.

COMPRESSOR VALVES

With weak valves, the frost line or sweat will disappear fast (less than 10 seconds) and the pressures will equalize in less than 20 seconds. The compressor head or dome will be too hot. The dome will be so hot that you will not be able to hold your hand on it. Bad valves will give the same pressure on both sides with the compressor running. An example will be 100 psi suction and 100 psi on the high side. The only way to repair this is to replace the compressor with a rebuilt compressor or a new compressor. Check the warranty on the compressor. Some compressors have a 5-year, 4-year or 1-year warranty.

A LOCKED ROTOR, BURNOUT, OR SHORT

Replace the compressor and liquid line drier. In the case of a burned out compressor, you must purge the system by hooking up a bottle of R 12 to the high side and blowing the refrigerant out the low side until you no longer detect the smell of the burnout. Before doing this purge, you have to cut out the burned out compressor. Purge the burnt system and install the new compressor with a suction drier. The suction drier will keep acid and sludge from entering in the new compressor on the suction side so that you do not have a second burnout.

Chapter 12
Compressor
Maintenance

Clean air-cooled condensers when they become dirty or clogged with dust. Clean water-cooled condensers if filled with mud or rust by using a brush or swab. Deline the water-cooled condenser with acid solution and acid pump.

Start up open drive compressors once a week for 2 minutes to get oil on the oil seal and lubricate this seal so that it will not crack and allow the refrigerant to leak out.

CHECK POINTS

Check the oil level in the crankcase of the compressor if you have sight glass. Check high and low side pressures. Too high a pressure and the oil will carbonize, leave the crank case, and oil log the condenser. Too low a pressure the oil will leave the crankcase and oil log the evaporator.

Excepting R11 systems, if the suction is 18-20″ Hg., you have an obstruction in the refrigeration system. If the refrigerant is short or low, the compressor will short cycle and run too hot a dome temperature. The refrigerant cools the windings of the compressor motor if it is semihermetic or hermetic. The air over the compressor dome helps cool the compressor motor also. The air that passes over the dome of the compressor motor is caused by the condenser fan motor as it brings air through the condenser and over the dome of the compressor, cooling both the condenser and compressor. With a refrigeration system low on refrigerant, the evaporator coil will be warm and have frost on the first two turns.

Too much refrigerant in the refrigeration system can occur during cool weather while charging a refrigeration system. With lower temperature and pressure, a refrigerant has more weight per fixed volume. When hot weather comes, the volume of the refrigerant expands and the head pressure becomes too high. This becomes hard on the compressor. If you have to charge a refrigeration system on a cool day, below 70 degrees Fahrenheit, block half the condenser with paper or cardboard and bring up the head pressure to 125 psi for R12 systems and 225 psi for R22 systems. Observe the sight glass; when there are no bubbles you have finished charging.

PRESSURE

The condenser pressure is the same as discharge or high-side pressure. I would like to propose a simple guide line that can be used to check condenser, discharge, or high-side pressures. Water-cooled condenser pressure will equal outside ambient air plus 20 degrees Fahrenheit. Look this temperature up in the temperature/saturated vapor pressure chart (Table 9-1) and you have a general idea of what the average condenser pressure should be.

Air-cooled condenser pressure will equal outside ambient air plus 25 degrees Fahrenheit. Look up this temperature in the temperature/ pressure chart (Table 9-1) and you will have a general idea of what the average air-cooled condenser pressure should be.

Air conditioning and refrigeration equipment have been designed for outside ambient air of 85 degrees F. Rule of thumb would be that water-cooled condenser pressure on the average day for an R22 system = 210 psi, R12 = 110 psi, and R11 = 5 psi for an average day. On an average day, an air-cooled condenser for an R22 system = 225 psi and an R12 system = 125 psi. With heavy heat load on initial start up on hot days, you might exceed the guideline high-side pressures. But as soon as demand is satisfied and you are maintaining load, you should be at guideline pressures or less.

You need to know condensing or discharge pressures because if you are charging a refrigeration system you can tell if you are overcharged by high-side pressures exceeding the guide lines. If there is trouble with the system—such as obstruction to refrigerant flow, dirty condenser or

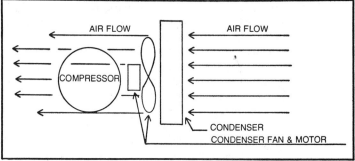

Fig. 12-1. A compressor/condenser air flow diagram.

evaporator, plugged air filters or driers, inoperative condenser fan or evaporator fan—these problems will all show up on the head pressure exceeding the guide line. When you are low on refrigerant because of charging or a leak, the head pressure will be too low and frost will appear on the first two turns of the evaporator coil at a particular stage.

Never tighten the springs under the compressor feet. The compressor must float on top of these springs. When the unit is new and shipped, the springs are tightened down. After installation the compressor must be made to float. Otherwise there will be too much vibration.

The air flow on a compressor/condenser package is through the condenser and over the dome of the compressor. See Fig. 12-1. The compressor is cooled by refrigerant within the system and air passing around and over its dome by the condenser fan motor.

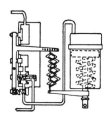

Chapter 13
Air Thermostats
& Air Controls

An air thermostat is a control that will regulate temperature to a fixed set point of a room, zone, or process cooling water by transmitting an air signal to a corresponding temperature pressure on the branch air line. The thermostat is supplied with 15 to 20 psi on its supply stud pipe. The branch pipe of the thermostat will have an air pressure (signal) that will control a valve, piston motor, or air pressure electric switch. There are two types of air thermostats which are direct acting and reverse acting. On the direct-acting thermostat, the temperature you are controlling is directly proportional to branch line air pressure. On the indirect- or reverse-acting thermostat, the temperature you are controlling is inversely proportional to branch line air pressure. The branch line is the tubing that leaves the thermostat and feeds the air signal to the valve, piston motor, or pressure electric switch. The branch air is the same as the control air to controlled device.

Figure 13-1 shows a typical air thermostat with supply air at 15-20 psi and control air to controlled device line. Note the control port where the bimetal element bleeds the right amount of air to give an air signal on the control air to controlled device line determined by temperature. The adjusting screw on all air thermostats is used to calibrate the thermostat to properly regulate and maintain temperature.

Figure 13-2 is a detailed view of a Johnson air thermostat. Note the sensitivity slider and sensitivity slider screw. When the sensitivity slider screw is in dead center of the feed-back arm, the temperature is most closely controlled in plus or minus of the set-point temperature of the dial.

Figure 13-3 shows the bimetal element A, the lever B, the dial and cam assembly C, and the calibration adjusting screw D of a Johnson air

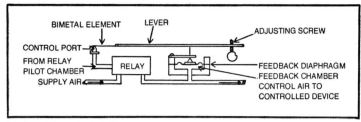

Fig. 13-1. A typical air thermostat.

thermostat. The letters RA stand for *reverse acting.* The letters DA stands for *direct acting.* If there is no printing on the bimetal element A, look for the black vertical lines and if the element has these lines, you will have a direct acting thermostat. When there are no vertical lines on the element, the thermostat is reverse acting.

In Fig. 13-4 you see two different types of Johnson air thermostats. The top side is direct acting. Note the letters DA on the lever that attaches to the black striped bimetal element. The bottom thermostat is reverse acting. Note the letters RA on the lever that attaches to the plane (non-striped) bimetal element. The letter D points to the calibration adjusting screw. The letter F points to the branch pressure test port.

Figure 13-5 shows a thermostat with two bimetal elements. Letter A is reverse acting calibration screw and letter B is direct acting calibration screw. It depends on the branch controls as to whether A or A1 is cooling or heating. Letter B is the branch pressure test port. Letter C points to the bimetal elements. Letter D is the temperature set point adjusting wheel.

Figure 13-6 shows another air thermostat. Letter A points to the calibration adjusting screw. Letter B points to the branch pressure test port. Letter C is the bimetal sensor element.

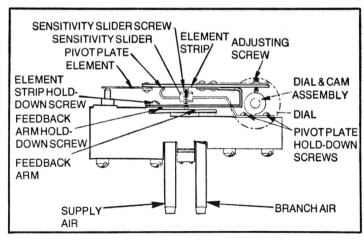

Fig. 13-2. A detailed view of a Johnson air thermostat.

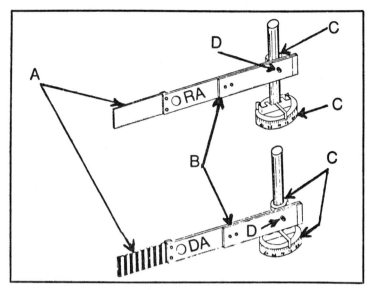

Fig. 13-3. Details of a Johnson air thermostat.

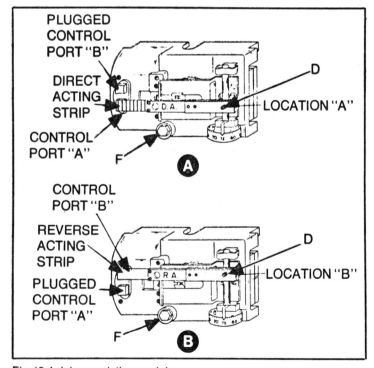

Fig. 13-4. Johnson air thermostats.

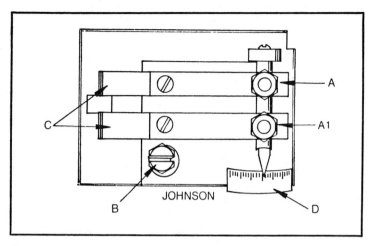

Fig. 13-5. A thermostat with two bimetal elements.

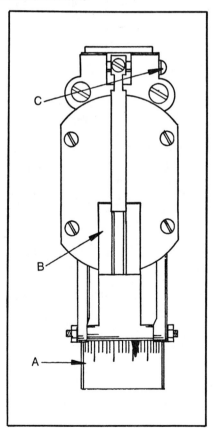

Fig. 13-6. An air thermostat.

Figure 13-7 shows a Johnson air thermostat with a branch pressure test port air gauge. This is a direct acting thermostat because of the letters DA printed on the lever and the black striped bimetal element. Note the calibration adjusting screw. The letter F is the branch pressure test port. This thermostat is calling for full heating because the branch pressure test port air gauge reads zero psi. Consult the following section on air thermostats so that you can read your branch pressure test port air gauge and know what the air thermostat is telling the system to do.

BRANCH PRESSURES FOR AIR THERMOSTATS (HOT AND COLD DECK SYSTEMS)

Direct Acting

    0   psi is maximum heat.

    5   psi is minimum heat.

    6   to 8 psi is a 50 percent mixture hot and 50 percent mixture cold air. With the demand satisfied, maintain room temperature.

    9   psi is minimum cooling.

  13 to 15 psi is maximum cooling.

Reverse Acting

    0   psi is maximum cooling.

    5   psi is minimum cooling.

    6   to 8 psi is a 50 percent mixture hot and 50 percent mixture cold air. With the demand satisfied, maintain room temperature.

    9   psi is minimum heating.

  13 to 15 psi is maximum heating.

To calibrate air thermostats of various manufacturers:

■ Obtain the accurate room temperature.

■ Set the thermostat at room temperature.

■ Insert branch pressure test port air gauge into branch pressure test port on your air thermostat.

■ Adjust air pressure on branch pressure test port air gauge to 8 psi by the calibration adjusting screw on the thermostat that controls tension on the bimetal strip. The air thermostat is a small regulator that changes air pressure to correspond to room temperature. The air thermostat has no differential. It is instant acting. At 8 psi branch pressure, you are maintaining the current room temperature.

■ If the thermostat has a stamp D A, this thermostat is direct acting. Use the direct acting pressure tables above. If the thermostat has a stamp R A, this thermostat is reverse acting and please use the reverse acting pressure tables above.

Figure 13-8 (A,B,C) are drawings I have made of the branch pressure test port air gauges used to check out and calibrate air thermostats. The little threaded nipple G or needle L is inserted in the branch pressure test port and will read branch or control air.

Figure 13-8(A) is a branch pressure test port air gauge that is used for Robertshaw thermostats. Letter A is a side view of a small simple air

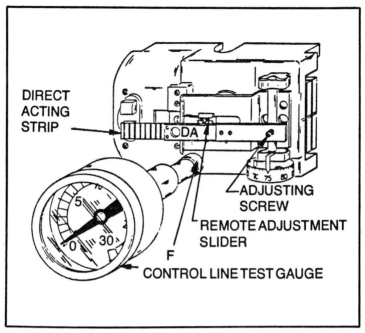

Fig. 13-7. A Johnson air thermostat with a branch pressure test port air gauge.

gauge from 0 - 30 psi. Letter B is a one-eighth inch pipe nipple off the air gauge. Letter C is a one-eighth inch female pipe thread to hose adapter. Letter D is a very small hose. Letter E is a small hose to male test port thread adapter. Letter F is a little rubber "O" ring. Letter G is male thread to branch pressure test port.

This air gauge and hardware is needed to read branch air pressures, control air, or the air signal that is put out by the air thermostat.

Figure 13-8(B) is a branch pressure test port air gauge that is used for Johnson air thermostats. Letter A is a side view of a simple air gauge 0—30 psi. Letter B is a one-eighth inch pipe nipple off the air gauge. Letter F is a little "O" ring. Letter G is male thread to branch pressure test port. Letter H is a one-eighth inch female pipe thread to male thread for a branch pressure test port adapter.

Figure 13-8(C) is a branch pressure test port air gauge that is used for Honeywell air thermostats. Letter A is a side view of a simple air gauge 0—30 psi. Letter B is a one-eighth inch pipe nipple off the air gauge. Letters J & K, is an adapter one-eighth female pipe thread to an air needle. Letter L is an air needle. On the Honeywell thermostat, insert the needle L of the branch pressure test port air gauge into the test port. When you are finished remove the needle. Over a period of years, the test port might leak. Use a small nail in the test port to stop the leak.

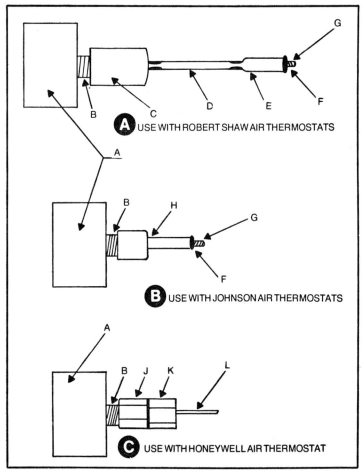

Fig. 13-8. Branch pressure test port air gauges are used to calibrate thermostats.

Figure 13-9 shows the same thermostats except the one to the right has the dial removed. Letter A points to the calibration adjusting screw. In this case, it is a little star wheel instead of a slotted head on the screw. Letter B points to the branch pressure test port. This port requires the branch pressure test port air gauge that is drawn on Fig. 13-8C with the needle L. Letter C points to the bimetal element.

Figure 13-10 is another air thermostat. Letter A is the calibration adjusting screw. Letter B is the branch pressure test port. You will need the branch pressure test port air gauge on Fig. 13-8(C) with the needle L for this thermostat. Letter C is the pointer used, to set the desired room temperature.

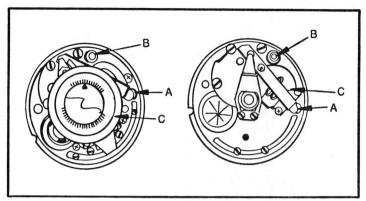

Fig. 13-9. Honeywell thermostats.

Figure 13-11 shows another thermostat. Letter A shows the calibration adjusting screw. Letter B shows the branch pressure test port. You will use the branch pressure test port air gauge that is shown on Fig. 13-8 (A) to calibrate this type of thermostat.

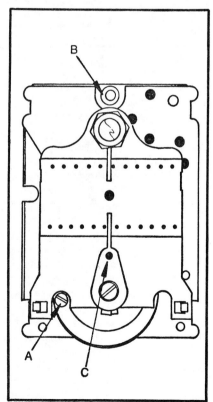

Fig. 13-10. A Honeywell air thermostat.

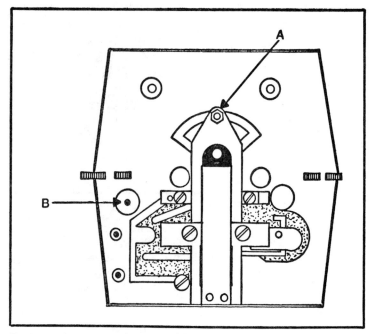

Fig. 13-11. A thermostat (courtesy of Robert Shaw).

Figure 13-12 shows another air thermostat. Letter A is the calibration adjusting screw. Letter B points to the branch pressure test port. Letter C points to the temperature sensor element.

Figure 13-13 shows another air thermostat. Letter A points to the nozzle assembly. This thermostat is calibrated by screwing in or out the nozzle assembly. There is no calibration adjusting screw. Letter B points to the branch pressure test port.

Figure 13-14 shows another air thermostat. Letter A points to the night calibration screw. Letter B points to the day calibration screw. Letter C points to the branch pressure test port.

I have shown you where the branch pressure test ports are located and the calibration adjusting screws are located. Follow my instructions on how to calibrate air thermostats and you will have success with all air thermostats. Consult the table for branch pressures for air thermostats for hot and cold deck systems and you will have a good reference point to start with. Using the branch pressure test port air gauges shown in Fig. 13-8, you will be able to see what the thermostat is doing. Set the thermostat at room temperature and make your corrections by turning the calibration screw in or out very slowly at one-tenth of a turn and wait for the branch pressure to stabilize. Set 8 psi on the branch pressure test port air gauge with the thermostat set at room temperature. Take a good thermometer with you to measure the room temperature.

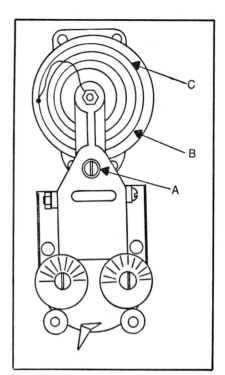

Fig. 13-12. An air thermostat.

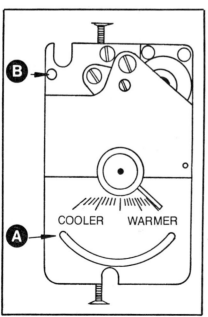

Fig. 13-13. An air thermostat.

COOLER   WARMER

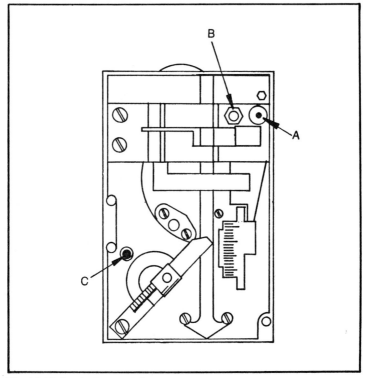

Fig. 13-14. An air thermostat.

TYPICAL APPLICATIONS OF AIR THERMOSTATS

Figure 13-15 covers basic air thermostat applications. The letters IV stand for *one way* air controlled valve. In this valve, the water can flow in one direction only—the direction of the arrow. The symbol 3V stands for three-way air controlled valve. The letter C stands for common. The letters NC stand for normally closed.

When the branch line that feeds the air valve is zero psi, the water will not flow from C to NC. The letters NO stand for normally open. When the branch line is zero psi, the water will flow from C to NO. Therefore,

Branch Line Pressure	Water Flow
0 psi	C to NO
8 psi	50% C to NO; 50% C to NC
16 psi	C to NC

The water can flow, depending on air pressure, from C to NO, C to NC, or C to NO and NC. The water can flow three ways and, hence, it is called a three-way air controlled valve.

The letter T stands for thermostat.

1—Tube thermostat has one tube connected to it.

2—Tube thermostat has two tubes connected to it.

3—Tube thermostat has three tubes connected to it.

The symbol XR stands for "restrictor." The 1-tube thermostat and transmitter need a restrictor for their operation. The restriction allows the 1-tube thermostat or transmitter to bleed off air and maintain a pressure on its feed that is less the supply air on the line side of the restrictor. The supply air is connected to the line side of the restrictor. The tubing on the other side of the restrictor is branch or control air.

The letters AM stand for air motor. The letters PE stand for pneumatic electric switch which is an air pressure operated switch.

The symbol $\frac{S}{20}$ means supply air at 20 psi. The symbol $\frac{D\text{-}N}{15\text{-}20}$ means supply air at 15 psi during day operation and 20 psi during night operation. The symbol $\frac{H\text{-}C}{15\text{-}20}$ means supply air at 15 psi for heating operation and 20 psi for cooling operation. The symbol ⊘ means a pneumatic gradual switch or variable remote air control pressure regulator (Fig. 13-16).

Figure 13-15(A) shows a thermostat that has a supply air of 20 psi and the control air or branch air is throttling a one-way air operated valve for chill water, hot water, process cooling water, or reheat water.

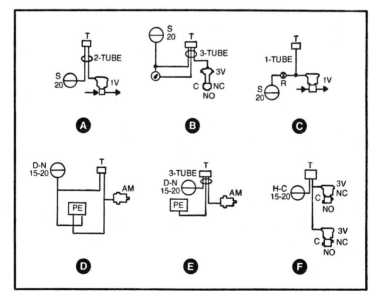

Fig. 13-15. Basic air thermostat applications.

Figure 13-15(B) shows a thermostat that has a supply air of 20 psi and the control air or branch air is throttling a three-way air operated valve for chill water, hot water, process cooling water or reheat water. In the air circuit, there is a pneumatic gradual switch that is fed by the 20 psi supply air and this switch can reset the thermostat and aid in control of the branch line.

Figure 13-15(C) shows a 1-tube thermostat 20 psi supply that air feeds the line side of restrictor XR. The load side of the restrictor feeds the thermostat which bleeds air off to drop pressure from 20 to 0 psi to correspond to my table for branch line pressures for thermostats. The restrictor XR maintains the supply air at 20 psi and will allow you to control and maintain pressures below 20 psi by the thermostat bleed-off. There is a one-way air controlled valve that tees off the restrictor. This valve will throttle chill water, process cooling water, hot water, or a reheat coil water. All the air line that is on the load side of restrictor XR is not only considered supply air, but serves as branch or control air at the same time.

Figure 13-15(D) shows two-tube thermostat that is fed by 15 psi of supply air during the day and 20 psi of supply air during the night. By changing the supply pressure to the thermostat, you change the set point and save energy dollars. During the daytime, the building is warmed or cooled for comfort air conditioning. During the night, the thermostat set point is changed by higher air supply pressure to reduce demand on the air-conditioning system and save dollars. The letters AM stand for air piston motor which works the dampers on hot and cold decks. The letters PE stand for pneumatic electric switch. This PE switch can control strip heaters, electric motors, or condenser units.

Figure 13-15(E) shows a three-tube thermostat. The air supply is 15 psi during the day and 20 psi during the night. The purpose of changing the psi on the supply air is to change the set point of the thermostat to conserve energy dollars.

There are two branch lines on this thermostat. One of the branch lines is reverse acting and the other is direct acting. The direct acting will feed the pneumatic electric switch and the reverse acting line will feed the air motor that is working the zone damper of a hot and cold deck system. The pneumatic electric switch can control motors, strip heaters, or condenser units.

Figure 13-15(F) shows a thermostat that has 15 psi supply air during the heating operation and 20 psi supply air during the cooling operation. The air pressure is changed in order to change the set point of the thermostat to conserve energy dollars. Changing air pressure can raise room temperature to 78 degrees for cooling and lower to 68 degrees for heating.

The branch line of the thermostat is feeding two air controlled three-way valves. This is a single zone fan system. The thermostat is direct acting. The hot water coil valve has a spring tension of 0 to 7 psi.

Fig. 13-16. A pneumatic gradual switch.

The chill water coil valve has a spring tension of 8 to 15 psi. The piping on the hot water valve is set so that you will have full flow to the coil in the fan unit at zero psi and bypass coil at 8 psi. The piping on the cooling coil is set so that you will have no flow to the CW coil in the fan unit from zero to 8 psi and full flow to the coil at 15 psi. The water is bypassed when there is no flow of water in the coil. In the fan unit, you will have one coil placed in front of the other. Normally, you will have the filters, the hot coil and the chill water or cooling coil and then the fan and duct system to the zone.

The heating coil is about one-third of the size of the cooling coil. This is because you have a higher differential in heating water than cooling water and, therefore, the size of the hot water coil is less than the cooling coil.

On a large building complex, install an inexpensive time clock that would control a solenoid valve which would feed an air regulator that would parallel the present regular that controls the building supply air. This system will give the $\frac{D-N}{15-20}$ or $\frac{H-C}{15-20}$ supply air. For the cost of about $70.00 for materials, you can change the set points of all the air controls of the building complex and the energy conservation in dollars would be quite large.

A large building complex could have more than 50= zones such as shown in Fig. 13-15(D,E,F). You would set the time clock to change the set points of the air controls at night and weekends when the building is not used. You can also raise cooling to 78 degrees F and lower heating to 68 degrees (Fig. 13-15(F).

HOT & COLD DECK SYSTEMS

The air thermostat is used with hot and cold deck systems or to throttle hot and chill water coils with air valves. A deck is a large square or rectangular duct. The air valves are basically three-way valves.

Figure 13-17 shows a hot and cold deck (duct) system with one air motor that has a branch line pressure of 13 to 15 psi. This air motor and damper system is controlled by a direct acting thermostat because it closes off the hot duct and opens the cold duct at 13 to 15 psi. If the air thermostat did the opposite—such as opening the hot deck (duct) and closing the cold duct at 13 to 15 psi—you would have a reverse acting thermostat.

The cold deck is a larger size deck between the hot and cold decks. If one duct is on top of the other, the hot deck is on top and the larger cold deck is on the bottom. Remember that the air thermostat can control an air motor with dampers, air valves on hot water coils or air valves on chill water coils. Air thermostats will transform temperature to air pressure and blend hot and cold duct air to satisfy room temperature setting with no time lag (differential). See Fig. 13-18.

In regulating air pressure, you will many times hear the air thermostat "hiss" air in its automatic bleed off. If you remove a DA air thermostat, you will have full heating. If you remove a RA air thermostat you will have full cooling. The moisture can be dried out in an air thermostat with Freon vapor. Oil and dirt can be blown out with Freon vapor. Then recalibrate the air thermostat.

Figure 13-19 shows a three-way air controlled valve. The three hub connections are threaded. Letter C stands for common. Letters NC stands for normally closed. Letters NO stands for normally open.

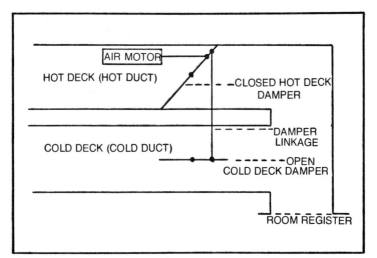

Fig. 13-17. A hot and cold duct system.

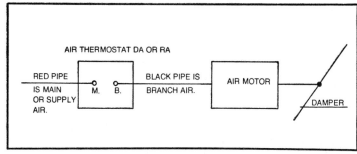

AIR THERMOSTAT DA OR RA

RED PIPE
IS MAIN
OR SUPPLY
AIR.

M.   B.

BLACK PIPE IS
BRANCH AIR.

AIR MOTOR

DAMPER

Fig. 3-18. Letter M is the main or supply air. Letter B is the branch air that goes to the air motor, air valve, or air switch.

To find common, cover one of the side hubs with your mouth and blow your breath into the valve. If you find your breath, circuit is from the side to the bottom hub—that side is common. The side hub that you cannot blow through is normally closed. When the branch or control air is at 8 psi, you have a 50 percent flow from C to NO and from C to NC. At 16 psi on branch or control air, the flow is from C to NC and the valve circuit of C to NO is closed. Whether the valve is direct acting or reverse acting on branch air depends on how it is planned in on the piping.

Figure 13-20 shows two three-way valves. The letter C stands for common. The letters NC stands for normally closed. The letters NO stands for normally open. The hubs are threaded hubs.

Figure 13-21 shows a quite large three-way valve. On all the large valves the hubs are flanged. You must use a gasket when you assemble a flanged hub. The flanges are nut and bolted together. The letter C stands for common. The letter NC stands for normally closed. The letters NO stand for normally open.

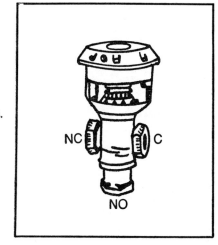

Fig. 13-19. Three-way valves.

NC

C

NO

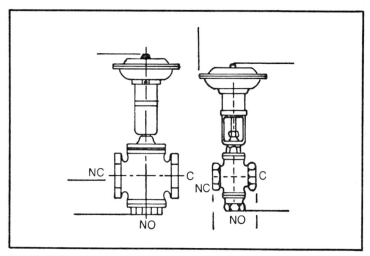

Fig. 13-20. Two three-way valves.

Figure 13-22A shows a one-way air controlled valve. These valves are small in size and throttle the water in one direction only. This can be used for heating or cooling.

Figure 13-22B shows a one-way air controlled valve. This can be used for cooling or heating. Mounted on the side of this valve is a pilot positioner which will cause the transmitter of thermostat to act very quickly on the valve. The purpose of a pilot positioner is to give quick response with no searching or delay on an air valve or air piston motor.

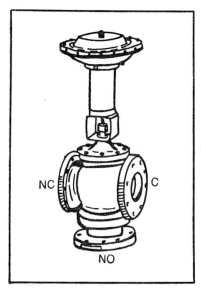

Fig. 13-21. A large three-way valve.

Fig. 13-22A. A one-way,
air-controlled valve.

Figure 13-23 shows a pilot positioner with an air piston motor. The pilot positioner is fed with a main air of 20 psi. The branch air is connected to the pilot positioner. The output of the pilot positioner is connected to the air device it is supposed to control. The purpose of the pilot positioner is to cause quick action of the air device it is controlling. The positioner will eliminate time delay and searching or hunting. On large piston motors or valves, the air volume is very large on the branch line so the pilot positioner voids this volume and takes over control of the air device for quick action with no hunting.

Figure 13-24 shows a pilot positioner. There is an adjusting screw to adjust response. The lever arm has a spring that attaches to the air motor or valve to trigger its action for quick response.

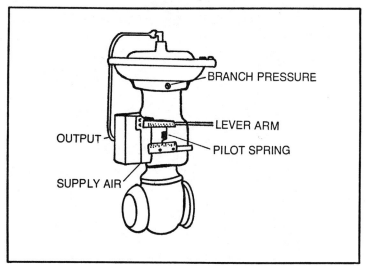

Fig. 13-22B. A one-way, air-controlled valve.

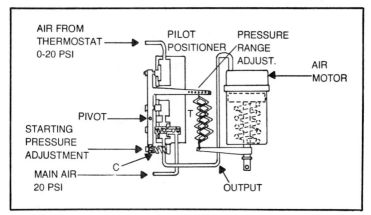

Fig. 13-23. A pilot positioner with an air piston motor.

PIPING APPLICATIONS FOR THREE-WAY AIR-CONTROLLED VALVES

In Fig. 13-25(A) there are three possibilities for this piping.

■ A could be first stage cooling and B could be second stage cooling. In this case, the easiest way to do this application is to use a spring of 2 to 7 psi for A valve and a spring of 8 to 14 psi for B valve. A direct-acting thermostat of the zone would cause A valve to come on first and be full open at 7 psi. At 8 psi to 14, B valve would come on and be full open at 15 psi.

■ We could have A be first stage heating and B be second stage heating. Use the 2 to 7 psi spring in A valve. Use the 8 to 14 psi spring in B valve. The thermostat would be reverse acting.

■ A coil is our primary cooling coil and B coil is used to further dehumidify the air.

In Fig. 13-25(B), the piping of the valves C and D is different from A and B. By using opposite thermostats, we are able to do the same three possibilities. Reverse acting is cooling and dehumidifying. Direct acting is heating.

In Fig. 13-26, NC is normally closed at zero psi branch pressure. NO is normally open at zero psi branch pressure. C stands for common. Figures 13-26(A) and 13-26(B) show how we can throttle a heat exchanger, boiler, water chiller, etc, by piping the valve two different ways. Figure 13-26(A) would have direct acting control. Figure 13-26(B) would be reverse acting control.

In Figs. 13-26 (C) and 13-26(D), we are piping the coil different using the same three-way valve. Figure 13-26(C) would be direct acting control for cooling and direct acting control for heating. Figure 13-26(D) is direct acting control for heating and indirect acting control for cooling.

Figure 13-27 shows a transmitter. A transmitter will change an air or liquid temperature into a corresponding directly proportional air pressure.

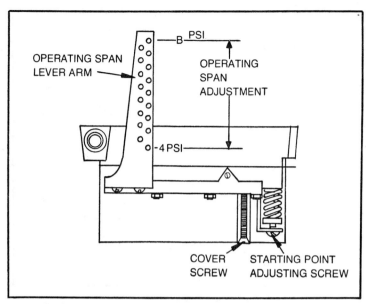

Fig. 13-24. A pilot positioner.

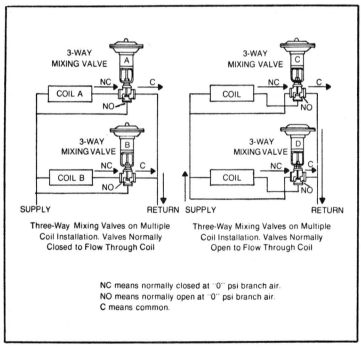

Fig. 13-25. Piping applications.

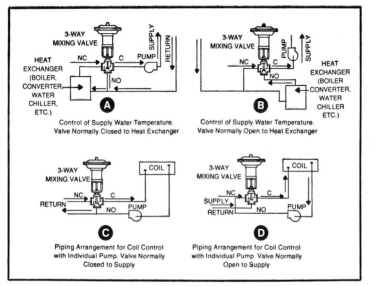

Fig. 13-26. Piping control.

The transmitter has no set point and therefore cannot control an air valve or piston motor. The transmitter is fed by one pipe which acts as supply as well as branch. The transmitter arrives at its temperature/pressure relationship by means of bleed-off of this pipe.

Figure 13-28 shows a restrictor. The restrictor is a small pinhole orifice that is placed in the air supply line that feeds the transmitter. This is the actual size of the restrictor and diameter of its pinhole orifice. This restrictor is used with plastic tubing. All air transmitters require a restrictor for their operation. The purpose of the restrictor is to permit the

Fig. 13-27. A transmitter.

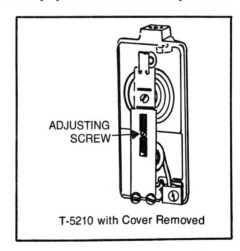

R-3710-2005 RED   .005 IN. ORIFICE
R-3710-2007 AQUA   .007 IN. ORIFICE

Straight coupling, in-line, with barbed fittings for ¼ in. O.D. polyethylene tubing (0.170 in. I.D.).
Material: Thermoplastic
Maximum pressure 25 psig (170 kPa)
Maximum temperature 180F (82C)

Fig. 13-28. A restrictor.

transmitter to establish and maintain pressures below supply air pressure by bleed off. Without the restrictor, the transmitter would try to bleed off the entire air control system and then nothing would function. The symbol X stands for restrictor. The arrow points to supply air flow.

Figure 13-29 shows a restrictor that can be fit inside a standard compression tubing connector. Actually, the restrictor is fit inside the one-fourth inch copper tubing and then the tubing is joined together by a standard compression connector. Letter A shows thermoplastic rubber O ring. Letter B is one-fourth O.D. polyethylene or copper tubing. Letter C is thermoplastic ferrule. Letter D is compression nut. The arrows point to the supply air flow.

Figure 13-30 shows a standard T restrictor. The supply air connects to letter A. The symbol X means restrictor and the arrow points to direction of supply air flow. Letter B can connect to the transmitter and letter C can connect to the air controller which will take the air signal of the

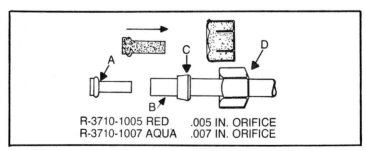

R-3710-1005 RED   .005 IN. ORIFICE
R-3710-1007 AQUA   .007 IN. ORIFICE

Fig. 13-29. A restrictor for inserting in one-quarter inch O.D. polyethylene or copper tubing. It is to be used in conjunction with various compression fittings—either in-line or at the controller. It includes a thermoplastic rubber "0" ring, zinc-plated compression nut, and thermoplastic ferrule. Maximum pressure is 25 psig (170 kPa) and maximum temperature is 180 F (82 C).

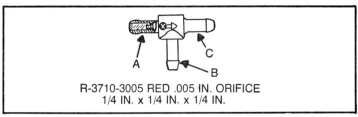

R-3710-3005 RED .005 IN. ORIFICE
1/4 IN. x 1/4 IN. x 1/4 IN.

Fig. 13-30. A "T" coupling, in-line with barbed fittings for one-fourth inch O.D. polyethylene tubing (0.170 inch I.D.). The material is thermoplastic and the maximum pressure is 25 psig (170 kPa). The maximum temperature is 180 F (82 C).

transmitter and do the actual control of air valve, piston motor or pneumatic/electric switch. If you wanted the actual temperature of the air signal of the transmitter, you could install a pneumatic thermometer gauge at letter C. Letters B and C are both fed from a restrictor inserted in letter A by the manufacturer. This is a cast thermoplastic part and thermoplastic tubing is squished over letters A, B, and C.

Figure 13-31 is a typical transmitter air circuit with tubing and restrictor (R), pneumatic/electric switch (A), and pneumatic thermometer gauge (B). Arrows have been drawn on the tubing to show the direction of supply air flow. Transmitter X is bleeding off the supply air to give a pressure of 11.3 psi on all the tubing that lies on the load side of restrictor R.

The letter S is the sensor bulb of the transmitter X. The air is bled off the transmitter at the bleed off port. The letter D of the pneumatic electric switch is the differential adjusting screw. Differential is the difference between cut-out and cut-in branch air pressure. On this switch, the cut-in is 2.5 psi and the cut-out is 12.5 psi. Therefore, there is a differential of 10 psi. By turning differential screw D, we can make the differential larger or smaller than 10 psi.

Letter R on the pneumatic/electric switch is the range adjusting screw. By turning this screw, we can raise or lower the set differential. This electric switch can make or break two separate electrical circuits. In this application, the electrical circuit is made at 2.5 psi branch pressure and opened at 12.5 psi branch pressure. This pneumatic/electric switch with relays is the controller and can control heating, cooling, fans, motors or solenoid valves.

You can use the pneumatic/electric switch to stage sequential operation of strip heaters, compressors, motors, solenoid valves and fans. This is most important as quality heating and cooling in all fields calls for the matching of horsepower, tonnage, or BTU to the heating or cooling load. See Tables 13-1 and 13-2.

Each stage requires a pneumatic/electric switch. The transmitter would be sensing the outside ambient air and supply signal to three pneumatic/electric switches. You can stage cooling in the same manner.

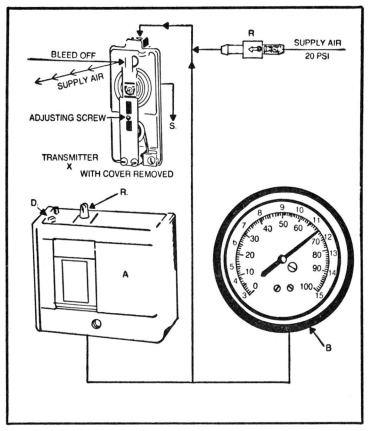

Fig. 13-31. A typical transmitter air circuit.

Table 13-1. Heating Examples.

First Stage is #1 Boiler cut in at 9 P.S.I. cut out at 11 P.S.I. #1 Boiler on the line.	= Minimum load + outside ambient air temperature minimum fall.
Second Stage is #2 Boiler cut in at 6 P.S.I. cut out at 8.5 P.S.I. #1 and #2 Boiler on the line.	= Average Load people are working. Getting cold from various ambient is 20° F. adding to the heating total load.
Third Stage is #3 Boiler cut in at 2.5 P.S.I. cut out at 5.5 P.S.I. #1, #2, and #3 Boilers on the line.	= Full heat load People are working. Cold from all sources, + outside ambient is −20 degrees F. The ambient is removing the boiler heat fast from the building.

Table 13-2. Cooling Examples.

First Stage #1 Compressor cut in at 6 P.S.I. cut out at 8 P.S.I. #1 Compressor on the Line	Minimum load early morning
Second Stage #2 Compressor cut in at 4 P.S.I. cut out at 6 P.S.I. #1 and #2 Compressors on the line	Medium Load 11:00 a.m. - people Various sources, + outside sun
Third Stage #3 Compressor cut in at 2 P.S.I. cut out at 4 P.S.I. #1, #2, & #3 Compressors on the line.	Full load 2:00 p.m. - people Various sources, + outside sun, full blaze

However, you would need a reverse acting control for the pneumatic/ electric switch. Transmitters are direct acting only. Do not forget, thermostat psi is not the same as transmitter psi. A thermostat has a set point but a transmitter does not have a set point. The temperature/ pressure table is only for transmitters.

On the pneumatic thermometer (Table 13-3) I have drawn in the air pressures that correspond to all transmitter sending (S) temperatures. Each 1 psi of branch air pressure equals approximately 8.5 degrees Fahrenheit. The thing I like about the pneumatic thermometer gauge is that I can tee it into the transmitter line anywhere and get an instant reading of temperature.

In Fig. 13-32, you see the air piston motor or piston operator. Piston operator A is unmounted and mounts on a plate using bolts into the threaded studs at letter C. Piston operator B is mounted on a plate at swivel point D. Letter E is the shaft that connects to a zone damper. The control or branch air connects at letter F.

Figure 13-33 shows the insides of an air piston motor or piston operator. The basic parts are the diaphragm, piston, spring, rod and piston motor body or shell. In normal position, the piston rod is pushed into the piston operator body by a stiff spring. On full stroke, the piston rod is extended by approximately 4 inches of travel by 16 to 20 psi branch pressure.

PNEUMATIC/HYDRAULIC LAW

Letter A = Force in direction of arrow
Letter B = Force in direction of arrow
Diameter of piston = 4″
See full stroke piston motor
Force = Pressure x area

Table 13-3. Branch Air Pressure in psi.

Branch Air Pressure in PSI	Pneumatic Thermometer Temperature in Fahrenheit
3	0°
4	8.5°
5	17.°
6	25.°
7	33.5°
8	41.5°
9	50.°
10	58.°
11	66.5°
12	75.°
13	83.°
14	91.5°
15	100.°

PRESSURE IN POUNDS

$$
\begin{aligned}
\text{Area} &= \text{square inches} \\
\text{Area of a circle} &= \pi R^2 \\
\pi &= 3.14 \\
\text{Radius} &= 1/2 \text{ diameter of a circle}
\end{aligned}
$$

Therefore,

$$
\begin{aligned}
\text{Area} &= 3.14\,(2)^2 \\
&= 3.14\,(4) \\
&= 12.56 \text{ inches}^2
\end{aligned}
$$

$$
\begin{aligned}
\text{Diameter of full stroke piston} &= 4'', \text{ radius} \\
\text{of full stroke piston} &= 2'' \\
\text{Branch Pressure} &= 20\,\text{psi}
\end{aligned}
$$

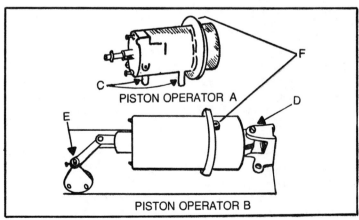

PISTON OPERATOR A

PISTON OPERATOR B

Fig. 13-32. An air-piston motor.

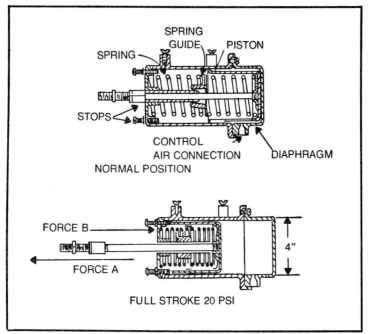

Fig. 13-33. An exploded view of an air-piston motor.

$$
\begin{aligned}
\text{Force A} &= \text{Pressure x area} \\
\text{Force A} &= \frac{20\,\text{Pounds}}{\text{Inches}} \times \frac{12.56 \times \text{inches}^2}{1} \\
\text{Force A} &= 20 \times 12.56 \\
\text{Force A} &= 251.20\,\text{pounds}
\end{aligned}
$$

The full stroke piston in the drawing without the spring, force B, would have a push of 251.20 pounds. If the spring was an 80-pound spring, then you would subtract the spring force B of 80 pounds from the piston force of 251.20 pounds to give an actual piston force of 171.20.

Therefore, the full stroke piston motor in operation can exert a force A on the damper of 171.20 pounds at a branch pressure of 20 psi. When the branch pressure is zero, the spring will move the damper back to normal position with a force of 80 psi. This force is needed because the piston motor is working two dampers (hot and cold deck) in two different directions.

Another possibility is to mount the piston motor vertically so that force A will act upward. Remove the spring from the piston motor. Secure a platform on the piston rod using the lock nuts. On top of the platform, place a weight so that weight equals platform weight 240 pounds. When you have the branch pressure at 20 psi, the piston will lift the 240 pounds the distance of the full stroke which is approximately 4 inches. Release the branch pressure very gradually and observe the platform push the piston to normal position. It is very interesting to see the power that a small piston can develop.

Figure 13-34 shows a steam humidifier with its reverse acting humidostat controller. The steam valve is normally closed with no branch pressure. The piping inside the fan has a lot of little spray jets to disperse the dry steam. The letters RA stand for return air. The letters OA stand for outside air.

The humidity of air that is heated several times becomes very low. In this case, we use steam to heat and humidify the building air. When branch psi is 0, the humidity is high and no steam is needed. At 8 psi branch pressure humidity is low and we are half open with the steam valve. When the branch pressure is 16 psi, the humidity is very low and the steam valve is full open to spray steam into the moving air stream inside the air handler to achieve 50 percent relative humidity for comfortable air-conditioning. Don't forget, too low humidity causes dry cracked skin, dry throats, static electricity and a lot of other problems.

In Fig. 13-35, the high pressure relief valve on the left side of the air tank is set to trip at 175 to 225 psi. This is a safety device. There is a little manual lever on the relief valve. I will pull the lever once a month and discharge air to see if it will discharge air and reset itself. The pressure switch cycles the motor. When the tank pressure drops below 95 psi the cut-in on the switch starts the motor. When the tank pressure builds up to 125 psi, the pressure switch opens to cut out the motor. Once a month I check the sight glass to see if we have the proper oil level on the compressor crank case. I also clean the compressor inlet air filter. The drain cock at the bottom of the air tank is opened once a month to remove water in the tank. Otherwise, water could get in the system and give trouble with the air controls.

The letter R in Fig. 13-35 shows an air regulator to drop overage tank pressure of 110 psi down to stable main air supply of 20 psi. The letter F shows a moisture and oil separator filter which will protect the air controls

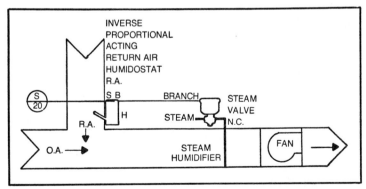

Fig. 13-34. A steam humidifier with a reverse acting humidostat controller.

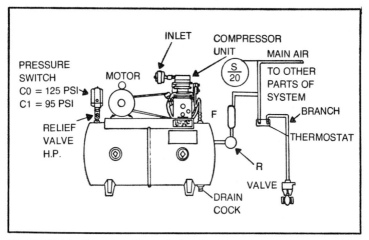

Fig. 13-35. A basic pneumatic control system.

from oil and moisture that could damage the air controls. On this filter, I change the desiccant every three months. Two other things I do monthly are to oil the electric motor and check the belt. If there is a lot of wear on the belt, I replace it.

Pneumatic control systems are made up of an air compressor to provide energy. The air lines are copper or thermoplastic. The supply air of 20 psi feeds the controlling device. Controlling devices are thermostats, humidity controllers, pressure controllers, relays and switches. Air lines leading from controlling devices to the controlled device are called branch lines. Air valves are called operators. Air motors are called acuators.

FLUIDIC RECEIVER (CONTROLLERS)

The purpose of a fluidic receiver-controller is to give closer control and quicker response and better regulation of zone temperature compared with air thermostats. The input signals to the fluidic controller are transmitter or remote set point. When a controller has a set point screw, the purpose of the set point screw is to control zone temperature. If the controller does not have a set point screw, then a remote set point feeds the controller to control room temperature. The set point can be reverse acting or direct acting. The gain screw of a controller sets the sensitivity and quick response of the branch air controlled device.

The ratio screw selects which input signal has authority and a relationship of one input to the other input. For example, if you wanted each signal to have equal authority, turn the ratio screw until No. 1.0 appears in the ratio window. If you wanted one control to be 75 percent of the signal strength of the other control, turn the ratio screw to No. .75 in the ratio window. When you get the controller new from the manufacturer, the signal amplification is 10 power. Because 1 psi transmitter output

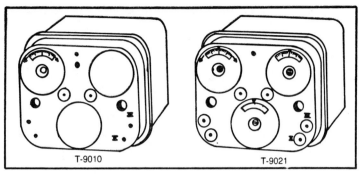

Fig. 13-36. Fluidic receiver-controllers.

signal equals approximately 8.5 degrees Fahrenheit, we will be using plus or minus .25 psi on either side of the set point of our fluidic receiver-controller to regulate zone temperature. The factory 10 power will automatically increase our controller output to 2½ psi, with an input signal change of .25 psi. A little signal change controls a large signal change.

Figure 13-36 shows two fluidic receiver-controllers on which you can see the gain, set point, and ratio windows and turning screws I have been talking about. Figure 13-37 shows the mounting base with fluidic receiver-controllers attached by two captive screws. The figures on the mounting base are a code for air connections. The air connections code is given in Table 13-4. This code is for Johnson Air Fluidic Receiver-Controllers.

The fluidic receiver-controller set point has been set for desired temperatures by turning the set point screw so that the top arrow lines up with a notch that will keep the room at the desired temperature. The transmitter is sending out a temperature/pressure signal to receiver-controller input. We are attempting to keep input signal to within one-fourth psi plus or minus of the set point by having an output signal

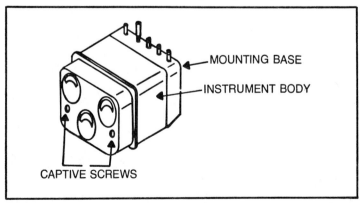

Fig. 13-37. A mounting base with fluidic receiver-controllers attached.

Table 13-4. Mounting Base Air Connections.

Connections CODE	Controlling-Receivers				
	T-9010 D.A.	T-9010 R.A.	T-9011	T-9020	T-9021
0	Output	Output	Output	Output	Output
III	Not Used	Not Used	Not Used	Master Transmitter	Master Transmitter
S	Supply Air	Supply Air	Supply Air	Supply Air	Supply Air
II	Control Transmitter	Remote Set Point	Control Transmitter	Control Transmitter	Control Transmitter
I	Remote Set Point	Control Transmitter	Not Used	Remote Set Point	Not Used

control a heating coil valve and piston damper operator that is 10 times larger than the input signal from the zone transmitter. This magnification of input signal gives excellent control and quick response to zone temperature.

Note the X restrictor in Fig. 13-38. Number II on receiver-controller stands for control transmitter. The letter S on receiver-controller stands for supply air. Letter O on the receiver-controller stands for output signal. Letter III is blanked off.

Figure 13-39 is similar to the prior drawing with two exceptions. In this drawing, we have a remote set point. The fluidic receiver-controller in this drawing might have a blank cover over the set point window and screw. The temperature of our zone A is controlled by a set point removed from the receiver-controller. The second exception is that in this drawing the output of the receiver-controller is providing air signal to piston damper operators on a hot and cold deck.

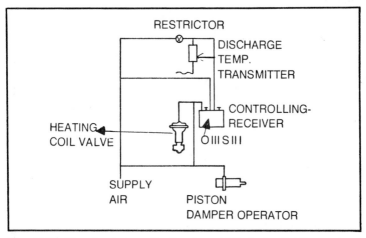

Fig. 13-38. The simplest application I could find of a fluidic receiver-controller.

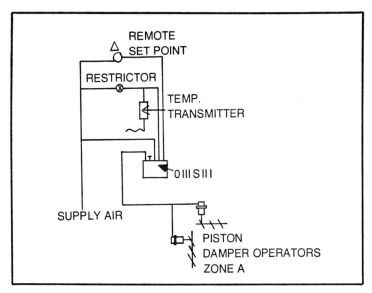

Fig. 13-39. Using a remote set point.

The X stands for restrictor. The letter S stands for supply air. The letter O stands for output signal. The letter I stands for remote set point, or control transmitter. The letters II stands for control transmitter. The letters III is blanked off.

In Fig. 13-40, you are controlling room temperature by two transmitters. The room temperature transmitter is in the room and is also

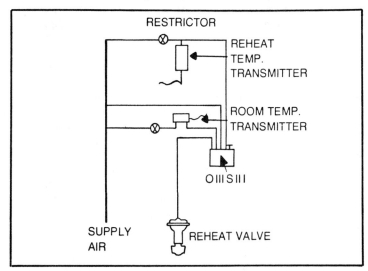

Fig. 13-40. Controlling room temperatures with two transmitters.

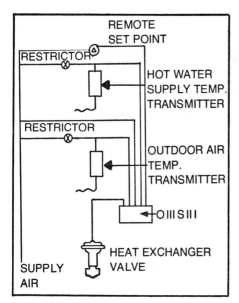

Fig. 13-41. Controlling the hot water of a boiler system.

known as the master transmitter. The reheat temperature transmitter is sensing temperature of the air leaving our reheat coil. The set point is on our fluidic receiver-controller and it maintains room temperature using the ratio setting on the controller and the air signal of the two transmitters. We can increase the gain setting to get more sensitivity and quicker response of controller to room temperature. The fluidic receiver-controller is providing output air signal to a reheat valve.

Letter I stands for remote set point or control transmitter. Letters II stand for control transmitter. Letters III stand for master transmitter. Letter S stands for supply air. Letter O stands for output signal. The symbol X stands for restrictor.

In Fig. 13-41 you are controlling the hot water of a boiler system. When the outside weather is hot, boiler water temperature is low or at minimum firing. If the outside air becomes cold, the boiler water is hot at maximum firing. During cold weather, you would run 180 degrees F. If the weather is quite warm, you can have boiler water at 100 degrees F and have good fuel savings.

This fluidic controller is provided with three input signals. The boiler system temperature is controlled by remote set point air signal at I on receiver-controller. The hot water and outdoor air transmitters provide air signals at II and III. The air signals at II and III are modified by the ratio screw and ratio number on the receiver-controller. By turning the ratio screw, one transmitter is given more authority than the other transmitter. At the 1-to-1 setting, the transmitters have equal authority.

The letter S is supply air. The letter O is output signal. The symbol X indicates restrictor.

Chapter 14
Refrigeration Testing & Charging Gauge Manifolds

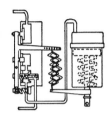

Figure 14-1 shows gauges mounted on a manifold. These gauges can measure liquid and vapor pressures. These gauges cannot be used for ammonia (R 717) because ammonia will eat up copper or brass. Ammonia refrigeration gauges are made of steel. The gauges in the field are the bourdon tube type and might have a small damper to keep the needle of the gauges from vibrating or quivering. If there is vibrating of the needle on the gauge so bad that you cannot read the pressure, you can correct this by shutting down the service entrance valve until the vibration quits or you can kink the hose that feeds the refrigeration gauge momentarily.

GAUGES

Gauge A of Fig. 14-1 is the suction, low side, or low-pressure side gauge. The outside band with the black numbers is used for psi pressure and the red numbers are used for vacuum pressure read out. Gauge A of Fig. 14-1 is a compound gauge. It can be used to read pressure in two different directions. The direction of arrow C of Fig. 14-1 is positive psi pressure. Gauge A of Fig. 14-1 has a maximum readible pressure of 120 psi or maximum pressure that can be applied of 250 psi.

The direction of arrow D of Fig. 14-1 in the red numbers of gauge A, is inches of mercury vacuum or negative pressure read out. The three inside bands on this gauge, from center to outside of R22, R12, R502, are temperature bands for these refrigerants and you can read the temperature of the refrigerant as well as the pressure at the same time within the refrigerant system.

Vapor temperature is directly proportional to vapor pressure. These bands correspond to the refrigerant/pressure tables in Chapter 9. A

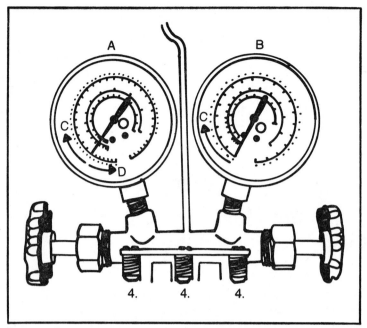

Fig. 14-1. Gauges mounted on a manifold.

compound gauge is used on the low side (or suction pressure) because many refrigeration systems such as R11 operate on the low side in 15 to 17 inches of mercury vacuum. If the unit is shut down and the evaporator is over 75 degrees Fahrenheit, you will read positive pressure. You can use the low-side gauge to tell you if you have a vacuum by reading 28 to 30 inches of mercury with sufficient vacuum pump running time. See the section on vacuum methods in Chapter 11.

The low-side gauge will tell you if you have a leak. After you have done the 24 hour vacuum or triple evacuation, shut the valves off on both sides of the gauges (valves E and F of Fig. 14-2) and wait one-quarter to one-half hour. If the gauges stay at the same vacuum pressure, there is an excellent chance your system is leak-free. I like to use the gauges to check leaks rather than soap bubbles, halide torch or an electronic leak detector. If you are worried about a small leak, leave the gauges on for two to four hours or over night. You can adjust gauge A and B to read 0 at atmospheric pressure of 14.7 psi by turning the adjustment screw inside of the crystal.

Gauge B of Fig. 14-1 is the high-side gauge. It is a positive pressure (psi) gauge (direction of arrow C) with a maximum pressure of 500 psi. The outer band reads gauge pressure. The inside bands from center out are R22, R12, and R502 and read temperatures directly for each of these refrigerants. Vapor temperature is directly proportional to vapor pressure.

There are three dummy one-fourth inch male flare fittings that are mounted on a plate on the manifold and I have placed a number 4 on these three dummy flares (Fig. 14-1). The refrigeration service man will attach his charging hoses to these No. 4 male flares in order to keep moisture, dirt, and dust from entering in the hoses, gauges or manifold. Always keep the hose attached to these No. 4 charging hose seal-off flares when the gauges are not in use.

CHARGING

The charging hoses that are connected to the gauges are open on one end and have an insert called an *indenter* on the other end. The indenter is used to depress the valve core of the *schraeder valve* so that you can enter the refrigeration system. A schraeder valve is a service entrance valve used in refrigeration and the valve core is the same as some auto tire tube schraeder valve cores. The flare that the schraeder screws into is one-fourth inch where the tire tube stem is much smaller in outside diameter. If you have a schraeder valve core that leaks on your refrigeration system, you can take one from a tire tube stem and use it. They are interchangeable.

Attach the open end of your charging hoses to No. 1, No. 2, and No. 3. If you have colored hoses: The blue hose is connected to the low side of the gauge manifold (No. 1). The white hose is connected to the center or common flare of the gauge manifold (No. 2). The red hose is connected to the high side of the gauge manifold (No. 3). This is standard connect up. If all hoses are the same color, you must trace the hoses each time you use the gauges. See Fig. 14-3.

The one-fourth inch male flare No. 1 is connected to the open end of the blue charging hose. The end of the charging hose that has the indenter is connected to the low side on the refrigeration system to the service entrance valve, schraeder valve or piercing valve. The one-fourth inch male flare No. 2 is connected to the open end of the white charging hose. The end of the charging hose that has the indenter is connected to the No. 4 dummy flare, charging bottle or vacuum pump, depending on what you are trying to do.

The one-fourth inch male flare No. 3 is connected to the open end of the red charging hose. The end of the charging hose that has the indenter is connected to the high side on the refrigeration system to the service entrance valve, schraeder valve or piercing valve.

Manifold valves E and F (Fig. 14-2) are closed on testing pressures. Manifold valves E and F are open on vacuum of the refrigeration system because I like to vacuum both sides of the system at the same time. A vacuum pump of 2 to 3 cfm is attached to the white charging hose (No. 2 flare on manifold). Follow a vacuum procedures outlined in Chapter 11.

On charging a refrigeration system, the refrigerant bottle is connected to the white charging hose which is connected to No. 2 flare on the gauge manifold. The white hose is purged of air by refrigerant vapor from the bottle with the white hose fitting loose on No. 2 flare for three seconds and then you tighten up the No. 2 flare hose connection. In charging, vapor is added on the low side of the system. On small systems, 7 tons or under, you charge on the low side. See system pressures in Chapter 9. The charging on large tonnage equipment (7 to 1000 plus tons) is done on the high side.

Using the same bottle hook up as a small system, valve off the high side of the refrigeration system. The system is started and you open manifold valve F and the liquid valve on the bottle. If the refrigerant bottle does not have a liquid valve, get the liquid by turning the bottle upside down. When the bottle is upright, vapor is what leaves the bottle. If the refrigerant container is over 30 pounds of refrigerant weight, it is usually called a *cylinder*.

The liquid refrigerant enters the system after the valve off and you keep charging until the sight glass is full or you can see that the liquid receiver is 75 to 80 percent filled. You might have to read the machine manuals the first time you liquid charge. These large machines are all different. You liquid charge large tonnage equipment because it would take too many hours to vapor charge. On small tonnage (7 tons and under), vapor charge on the low side only.

To equalize the pressures in a system:

■ Turn the system off.

Fig. 14-2. Gauges mounted on a manifold.

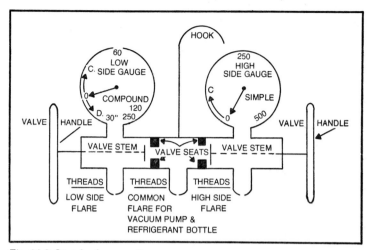

Fig. 14-3. Standard connections.

■ Connect the blue charging hose to the low side and the red charging hose to high side. Hoses are connected to the manifold on the other end.

■ Connect the white charging hose, which is on the No. 2 flare on the manifold, to the No. 4 dummy flare. This No. 4 is the charging hose seal-off flare.

■ Open the manifold valves No. E and No. F. You have now bridged the low side to the high side and the pressures will come down and equalize fast. This little trick will let you start up the machine under partial pressure. If the head pressure is too high too fast, you can stop the machine and equalize pressure before damage is done to the valves.

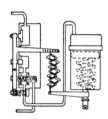

Chapter 15
Air Filters

The purpose of air filters is to remove atmospheric impurities. Some of these impurities are dirt, dust, smoke, smog, fumes, odors, and pollen. The air filter is normally installed in racks in the return air of a fan system before the suction of the supply fan. The air filter will help keep the fan motor windings clean. If the fan motor becomes quite dirty, the motor will not be able to rid itself of the internal heat that it develops in doing the work of turning the fan. The excess heat, in time, will cause dielectric breakdown and a motor burnout. This happens because all these atmospheric impurities are a form of insulation. By keeping the fan clean, we can solve the problem of dirt buildup on the blades which will upset the balance of the fan blades and cause severe vibration, metal fatigue, wear and tear, plus noise to the area being served by the fan system.

A good air filter system will protect from pollution the environment served by the fan network. However, in order to protect the indoor environment we must use special kinds of filters that will be covered in this chapter. The ducts are kept clean on the inside by a good filter system. Most important is that the cooling and heating coils be kept as clean as possible.

One micron is .000001 of a meter. A particle of matter 1 micron in diameter is .0000394 inches in size. I have labelled the percentage of efficiency of filtration of different types of particulate matter. For example, 35 percent efficiency filtration will handle fly ash, dust, descending impurities, molds, fog, pollen and industrial dust.

To be effective in filtering bacteria, the filter will have to be 55 percent efficient. To be effective in filtering fumes, it must be 85 percent efficient. To be effective on suspended impurities, oil smoke and tobacco smoke, you will have to have 95 percent efficiency in the filter network. Less than 10 microns is not visible to the unaided eye. Consult Fig. 15-1.

COOLING COILS

If the cooling coil becomes clogged with excess dirt buildup and the cooling coil is a direct expansion coil, you will find high head pressure on the compressor and the refrigeration unit will shut down automatically on the high head pressure cutout switch. If the switch does not function or is tampered with, the compressor will burn out. An old saying goes that if you don't change filters, you will change compressors, is true as ever.

A direct expansion coil can be called an evaporator and is fed by a thermostatic expansion valve, cap tubes or other refrigeration pressure device. Dirt insulates the cooling coil and the coil cannot function. The heating coil is very similar to the cooling coil. The heating and cooling coils of an air conditioning system are heat exchangers and must have fins or plates that have minimum dust and dirt so that the air flow through the coils will not be insulated from the fin or plate area and that the exchange of heat is completed. Excess dirt will limit the supply air volume to the inner environment that is served by the air conditioning system. The heat load of the environment will overtake the temperature of the supply air and the rooms will be too hot or cold.

What do you do if you find that you have a very dirty coil. First, remove the filters that are in front of the coil or coils. Second, obtain compressed air, nitrogen or carbon dioxide. I like the nitrogen or carbon dioxide, because it comes in a bottle, is portable, and a regulator can be installed on the bottle with a hose and nozzle or just a plain open ended hose. Set the regulator at 100 psi to start and direct the gas so that it travels parallel to the plate or fins. Do not be at an angle to the fins as you will bend them and damage the coil. Work the hose slowly up and down the fins. Go up vertically with the fins and then cross about 1 inch. Come down and cross over 1 inch and go up and keep repeating until you have covered the entire coil.

If blowing the coils with compressed gas does not work, next try to spray the coil with soap and water. This will require a pump and hoses with the same technique as with the compressed gas. Before starting, check to see if there is an electric strip heater on either side of the coil or electric controls or wiring near the coil. The power must be turned off and the controls must be covered by plastic or waterproof tarp. Use the same up and down motion or run parallel to the coil with the soap and water solution, with the 1 inch cross-over. Let the soap and water soak for 20 minutes before you rinse with plain water.

I found an excellent solution for this spray process is to take a water pail, fill it 80 percent with water, put in 1 cup tincture of green soap, 1 cup Spic & Span, and gently stir. If additional action is needed, add one-half cup ammonia. Gently stir as you get a soap foam on top of the pail. Use a small

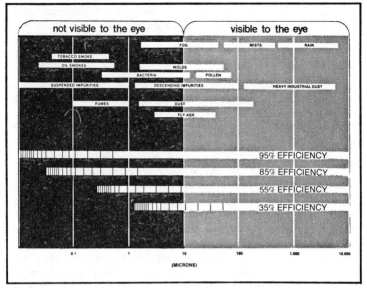

Fig. 15-1. Particle size and filter efficiency.

pump, electric or gas, and place a hose on the discharge of the pump, the same size as a refrigeration charging hose. On the discharge of the pump, you must reduce the hose size so that you can get a high pressure spray or stream of solution into the coil and through the coil. Try to have a self-priming pump so there is less hassle.

If field conditions and time do not permit the soap and water approach, try this: Secure the largest diameter garden hose you can get and hook it up to a hose bib. Screw onto the garden hose a spray nozzle that has a trigger on it. Try a spray and stream approach, using the trigger and cover the coil with the same procedure as outlined earlier.

When the gas blow through, soap and water, and water spray fail, there is only one last resort. Secure a portable steam cleaner. Direct a pressure stream into the coil and attempt to spray steam parallel. You will bend the fins a slight bit, but this can be corrected with a fin comb.

Attempt to sponge up all excess water. Check to see that wiring is dried off, the power is turned off of the strip heater, and that controls are dry. Close up the fan room and start the fan to dry out the coils. Use no heating or cooling—just the fan. Run the fan one-half hour. Next, place clean filters back in the filter racks, turn on the strip heater circuit and energize wiring and controls if the area is dry. Place the unit in service.

When the coils become plugged due to deterioration, alkali, minerals, decomposed aluminum or copper, try the following steps. On large fans, you may have to cut an access door to enter the coil area. Set up a good work light inside the fan. Place a canvas tarp on the insulation that is on the floor of the fan. Try not to break the black felt layer of the insulation. Get a small mask (disposable) and place it over your nose. The tools you will need are paint scrapers and putty knives. You will find that the decomposed front of the coil can be scraped away to the first row of copper tubes. The rows of copper tubes are staggered and it is not possible to punch or rod a refrigeration coil. After you have removed the fins to the first row, take a look and see if there still is alkali, minerals, dust, and decomposition between the fins.

You will probably have to scrape around the first row of tubes and bow the tubes out and scrape to row No. 2. The fins on standard refrigeration coils are not welded or bonded or soldered. It is a friction fit. When the coil was made, the copper tubes were inserted in the holes that were punched in the side of the fins. Next, a steel ball slightly larger than the inside diameter of the copper tube was forced through the tube by air pressure. The copper tube was expanded and the aluminum fin was beveled on to the copper tube in a tight friction fit.

In order to have eight to 11 fins to the inch, as most refrigeration coils are, a jig is used. In practice, I have had to go as deep as three rows before all the minerals, alkali, and decomposition was scraped away and all I could see between the fin was open dark space. Putting a light on the other side of the coil will not tell you anything because of the staggering of the vertical rows of copper tubing that carries the water or refrigerant inside. Don't worry about supporting the loose copper tubing that is not supported anymore by a fin. It is thick-walled copper and will not let go. It will also pick up some condensate and do a small amount of heat exchange.

Get your brooms and dust pan and remove all the debris that you scraped away. Next, you must vacuum with a vacuum cleaner the fins and canvas that you have on the floor. Do not vacuum the insulation that is inside the machine. If you do, you will find that the vacuum will remove the black felt. The purpose of the felt is to keep the fiberglass, asbestos, or other insulating material from becoming picked up by the air stream and polluting the fan supply environment with small particles of asbestos, fiberglass, or other irritant materials to eyes, throat, and lungs of the people working and living in these rooms.

Before you start the fan, remove the canvas tarp and be careful not to tear up the felt on the floor. If you find that the felt is gone or all broken up and that the insulation is in bad shape, you can lay down new insulation. I like to use arabal lagging adhesive made by Borden Chemical Company and thin it out with water and paint it on the old insulation with a paint brush. You might also purchase a bag of wheat flour from the grocery store and mix it with water and make wheat paste of a thin consistency. Paint the wheat paste on. Now put the trap door back on the fan.

You have one more problem to attend to before we start the fan. Cover all the furniture with plastic, paper, or tarps because when you start the fan you are down stream of the filters with the work and the residual dust will come out of the registers. I like to run the fan without air conditioning for at least two hours to eliminate the dust not removed with the vacuum cleaner.

Next, remove the dusty paper or plastic that is on the furniture. You can remove 15-20 percent of the coil fin area and still get by. If need be, you can run a lower temperature chill water to the coil or start the cooling sooner. *Do not* adjust super heat on the thermostatic expansion valve if this coil is direct expansion coil. This operation on a large fan will take two men one good day, but could save the corporation that owns the building over $30,000. Bad coils are easy to remove because you can take them out in pieces. However, to install a new coil, say about 28 feet long by 4½ feet high by 14-16 inches thick, is going to require many different trades, time, effort and money.

How do you know when you have a very dirty, particularly plugged, coil or decomposed coil. You will find the room too warm or too cold with not enough supply air to overtake heat load and control rooms temperatures. Before you jump to conclusions, check air filters, fixed dampers in the system, and hot and cold deck zone dampers. These three situations will give the same trouble as a dirty, partially plugged coil or decomposed coil.

FLAT DISPOSABLE FILTER

You can purchase flat disposable filters at hardware stores and many retail stores. The frame is of cardboard, with pads that are held in place by a grid work. The pad media is usually fiberglass or other man-made media. Normally, there is an arrow that will show the direction of air flow. If you hold this type of filter to the sunlight, you will find that there is a fine oil barrier sprayed on one side. The purpose of the oil barrier is to catch and trap particles that are airborne and hit the barrier.

Normally you will find the dimensions to be a little undersized when you measure these filters and compare them to the printed label. For example, a 24″ x 24″ x 1″ will measure 23½″ x 23½″ x ⅞. From time to time you will encounter a fan system that does not have standard size filters. This might be because many older fans used media pads that the utility man used to cut and place inside of steel frames that had a grid network that clipped in or bolted in.

If you are working with a filter with a steel frame, remember the oil barrier. The oil barrier faces away from the fan or coils, whichever is first. I used to remove these from the fans, wash them out with water, and set them beside the wall to dry. Then I would use a can of spray stickum and spray the barrier on the front. These filters were rusty, very messy, and used up a lot of labor units. I have not seen them in years and I surely do not miss them. Although you could use them over and over, no money was saved because of the heavy labor drain.

When the frame and pad system and metal filters were discarded and replaced by throw away disposable, this situation gave us the oddball sizes. Some manufacturers used to make a u-frame filter. This was a media placed in a gridwork that wrapped around the fan. These u-frame filters were replaced by a fixed filter rack that might not be standard in size.

American Air Filter Company makes a line of air conditioning equipment called Herman Nelson. Many of these machines used a roll-type filter. There was a spool on either side of the return air of this unit. One spool, called a "take-up" spool, was turned by a crank and worm gear. The other spool contained the supply media. The trouble came when the worm gearing froze and the crank broke. Then it took two men, one man on either spool, to change the filter media. One man would unwind the supply media and the other would wind up the dirty media. I hope to never see this affair again.

What do you do when you come to a non-standard filter or a filter frame for which you do not have the proper filter for replacement? Measure the filter rack and see if you have a filter that has the same size as the length or width dimension. Next, take scissors and cut the other dimension to size. Take a coat hanger and straighten it out and lace it through the media and paper framework near your cut and size it to the paper frame. The purpose of the steel wire is to prevent the filter from collapsing. If you have sheet metal you could make a sheet metal end piece and clamp it on to the cardboard frame. Many customers are unwilling to pay for new filter framework or the labor units to get the proper sizes, if available. The success of any operation is to give the customer what they want.

When do you change disposable filters? A general rule on air conditioning and heating is to change them every 60 to 90 days. If you are in an area that has a lot of dust and pollution, use the following method. Open up the fan and look to see if there is a dust buildup on the face or oil barrier side. Pull one filter from the rack and see if there is a bleed-through and discoloration on the underside that is next to the fan or coil. Also look to see if the filter has collapsed or the media has opened up. When any of the above conditions develop, change out the filter immediately.

Many firms stock filters and use their personnel to install the stocked filters, labor time permitting. Other firms will use a filter service by many firms that do this work. The idea is that they come every two or three months automatically and service the filters at a minimum labor charge. They make their money on filter and labor and a purchase order eliminates storage, individual buying by maintenance personnel, and minimizes record keeping.

Purchase one grease gun with hose and a 90-degree fitting, one oil can, and the belts needed for the fans. When the filter man comes, he will install the filters, plus grease bearings every six months and oil bearings every six months. When you grease the bearings, pump the grease from the gun until a small ribbon comes out of the bearing. Use four to six drops of oil in the oil bearing. Fill grease and oil cups to their mark. One important detail is to keep a written record of when the filters are changed for the various fans.

When the filters are being changed, check out the belts of the supply fan and change them out if they are frayed, worn or torn, or show cracks in them. Use air conditioning contractors for major troubles. Their labor units are too high for filter work, changing belts, oil and greasing fans, and motors and water chemical work.

The efficiency of the flat disposable filters runs 15-20 percent. The efficiency of 30/30 or pleated filters is 30 percent. These type filters have up to seven times the area of the flat disposable and the service time may be extended. However, use the same vision check-outs as with the flat disposable filters when changing the filter.

Note that the wire on the back side of this filter gives it a great deal more grid support than the flat disposable filter. Pleated filters can be as thick as 12 inches and depending on the media, the efficiency can run as high as 55 percent to 95 percent. With large surface area, the air volume will be reduced to give longer service life and the air resistance of the higher efficiency filters will be less. As a consequence, a savings will come because of less work by the fan motor.

These filters could be used in clean rooms and laboratories and can help in allergies or hay fever. They would protect priceless oil paintings and art work. A high performance air filter can have up to 95 percent efficiency, depending on media and size.

THE MAGNAMEDIA OR ABSOLUTE AIR FILTER

The efficiency of this filter will run 95 to 99.9 percent. This is a honeycomb-type filter with aluminum, cardboard or asbestos separators and the media is fiberglass. The temperature that the filter can withstand can be as high as 800 degrees F and the humidity can be as high as 100 percent in special cases.

The clean magnamedia or absolute air filters will have a manometer reading of .75″ water column. This filter is dirty and should be changed when the manometer reads 1.50″ water column.

ACTIVATED CARBON FILTERS

The purpose of the activated carbon filter is to remove objectionable odors from the air conditioning environment and purify the air so it will smell clean. Most of these filters are a screen that surrounds a steel frame with many compartments filled with carbon.

When you find a gray color appearing on the screen or the carbon turns gray, it is time to change out the filter. Although most of the carbon filters are throw-away panels, you will find some expensive carbon filters where you can remove the old gray colored carbon and replace it. A system of carbon filters can be made up of many panels in series that look like the deep pleated filters.

THE ELECTRIC FILTER

This is also known as the electrostatic precipitator. Many fans will have this type of intermediate filter. It will consist of two parts—many grid wires and many plates. The plates will be vertical from top to bottom on the fan and spaced one-half inch apart the width of the fan. The width of the plates will normally be 10 inches plus. The grid wires are located approximately 2 inches ahead of the plates and between the plates vertical and suspended by high voltage insulators. The plates and grids are either aluminum or stainless steel.

In operation, a high voltage over 10,000 volts plus dc is placed on all the grid wires. The plates are normally at ground potential. When the air stream passes the grid wires, all the particles become plus charged because the grid has a magnetic field or electric flux around it. Many large filters will have a small hum noise on the grid and this magnetic field is known as ozone and has a sweet, clean smell. Unlike charged particles attract each other and stick to each other; like charged particles repel each other. After the dust and pollution passes the grid and becomes plus charged, we find that when the particles of dust travel between the plates at negative ground potential. The particles are sucked on to the plates and hold on like little magnets.

These electric filters normally have a sequence clock. When they become filled with particulate matter, the clock will shut down the air conditioning and fan. Ahead of the filter you will find shower heads that will come on by the clock and wash down the plates and grids. When the grids and plates are clean, the clock will shut off the shower water and start the fan. It takes a very short time for the plates and grids to dry out and when this happens the air conditioning unit will start and the high voltage power supply will energize the grids. Then you are back on the line with the electric filter. The dust and other particulate matter leaves the electric filter by way of a drain pan and drain pipe and is disposed.

Behind the electric filter plates you will find many metal filters in a rack and these are called *moisture eliminators*. Their purpose is to trap the moisture after it leaves the electric filter and keep the after filters in the fan system dry.

These electro-static precipitator filters are used with gold and silver smelters and power plants. Instead of the water cycle, you will find rapping hammers operated by compressed air that will knock the particulate matter from the plates into a hopper. Under the hopper is a sliding door that operates like a gate valve. Attached to the opening around the door is a large barrel on a clamp-type barrel ring. By collecting precious metals, the electro-static filter will pay for itself many times over. Many years ago, I worked at an electric generating power plant on a smog program. The electro-static filter did two operations. It can reduce smog on bunker fuel fired boilers and plus we found that the particulate matter in the barrels was very rich in ammonia and would make excellent fertilizer.

THE SCRUBBER

In laboratories, you will find a device called a *scrubber* and this will clean up the air of acid fumes, ammonia vapor, airborne poisons and foul

123

odors. What you have is a large plastic rectangular box with a bottom drain. Inside this box are different plastic media filters, one behind the other. Water is sprayed on the front of the filters and also spilled over the top. The different configurations of the media will impinge the dust and the water will act as a barrier as well as rejuvenator and cleaner. You will find the scrubber filter used where the air is so polluted that the activated charcoal could not keep up or the down time is more than the run time.

In the air conditioning field, you will find single-stage filtration, two-stage filtration, and three-stage filtration. In single-stage filtration, you will find either flat disposable, pleated, or renewable steel mesh filters in the filter rack on the return air of the fan system. In two-stage filtration, you will find the first filter rack called the pre-filter to contain flat disposable or pleated filters. The purpose of the pre-filter section is to trap the larger and medium particulate matter. In doing this you extend the life of the expensive final filters that are in a filter rack behind the pre-filters.

The final filters can be carbon, a pleated filter that has the extra depth of media to boost the efficiency to 55 percent, 85 percent, 95 percent of high-performance filters that range 55 percent, 85 percent and 95 percent, depending on media. In three-stage filtration, the pre-filter section will contain the flat disposable or pleated filters.

The intermediate filters will be carbon, electro-static precipitation with the extra deep pleated in the 55 percent, 85 percent or 95 percent range, or the high performance filters in the 55 percent, 85 percent or 95 percent range. The intermediate filters will protect the very expensive after filter from loading up with dust and extend its life for a very long time. The after filter will either be an absolute or magnamedia, depending on the manufacturer, with 99.9 percent efficiency.

Three-stage filtration will handle all clean rooms. If you have problems with pollen, allergies, asthma, or smog, try to get a job in a clean room. Not only is the air clean, but the humidity is controlled in the 32-36 percent relative humidity with a temperature of 71 degrees to 72 degrees Fahrenheit. The room has one-tenth inch water column pressure between the inside and outside of the room.

The slight positive pressure in the room is to keep dust from migrating into the room from the cracks around the door, the ceiling or tiles. It is the ultimate in air conditioning.

With two-stage filtration, the manometer reading across both filters will be around 2 inches of water column. Three-stage filtration, the manometer reading across all three filters will be around 3 inches of water column. With expensive filters, like absolute, high performance, deep pleated, magnamedia, it is not possible to tell by vision the condition of the filter bank. We use a device called a manometer and compute the static differential pressure. Static pressure is the pressure or force that the air pushes against the inside walls measured in inches of water column. Differential pressure is the difference between the static pressures on either side of an air filter.

There are two types of manometers made. One is a mechanical gauge that has a diaphragm that moves a needle that points to numbers on a face plate. This is a non-linear gauge. If I had 10 of these mechanical gauges, new, all the same model, calibrated from the factory, and I placed them in service, they would all read slightly different static pressures if used on the same fan at the same location. This is what is meant by non-linear. The Bourdon gauges, whether used for vapor, gas or liquid, are all non-linear.

The u-tube gauge manometer is the simplest, lowest cost, most accurate, and it is linear. This manometer is nothing more than a tube that can be any diameter in the shape of a "U." One-half way you have the reference number "O" that extends from one side to the other side of the "U". Then you have two wide rulers reading in inches that extend in either direction from the "O" to a length of 4 inches. The fluid inside this U-tube is water with a little vegetable coloring in it. There are two little corks on either end of the "U" to keep the water from running out. The distance between the sides of the "U" does not matter.

Next, you take a rubber hose about 3 feet long of small diameter (makes no difference what the diameter is) and attach it to one end of the U-tube. Keep the tube upright and have the corks on the end removed. Water reference level is at "0" on both sides of the "U."

Get a cork stop with a little hole in it, the same size as the hose that is attached to the U-tube and shove the hose through the hole of the cork so there is a little extension (say one-fourth inch) of the hose beyond the cork. The cork will have a taper and the narrow diameter will be on the hose one-fourth inch extension. Next, take an awl and make a hole the same size as the narrow taper of the cork in the fan sheet metal about 4 inches to 12 inches on either side of the filter rack.

Insert the cork with the hose in one of the holes in the fan sheet metal and use a spare cork to stop the other sheet metal hole. Note the static pressure (inches of water column) and write it down on paper. Repeat the same procedure on the other sheet metal hole and cork up the hole you just tested. Note the static pressure and write it down. Subtract the smaller water column number from the larger water column number and you have the differential.

If you were to use two hoses and cork stops and attach one hose to one of the sheet metal holes, and the other hose to the other sheet metal hole, the U-tube manometer would read the differential pressure. Hold the U-tube vertical and make sure the water level, with no hoses attached, is at "0."

This test is the most accurate. Every time you check the water level to see that it is at "0" with no hoses means you have calibrated your gauge. This gauge is linear. Take any u-tube gauge with any tube diameter, hose diameter, cork diameter—do your water level calibration to "0"—and they will all read exactly the same water column number.

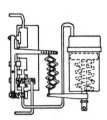

Chapter 16
Water Chemistry &
the Air Conditioning Field

Because you are dealing with large volumns of water in the air conditioning and heating fields, there are certain steps you must take to maintain and safeguard equipment. In this chapter, I will cover pH control, biocides, herbicides, phosphate, chromate, bleed off, and other information that may be of help.

When condenser water from a large water-cooled system is heated by the condenser that transfers the heat of the machine refrigerant to the water that flows inside the tubes of a shell and tube condenser— and then this water and heat is taken outside to a water tower, spray pond or waterfall—an interesting situation happens. The heat is always transferring itself to a cooler medium. The outside air is the cooler medium. When the heat leaves the water, it will take a lot of water vapor with it by evaporation. This leaves minerals, alkali, dissolved solids, and salts remaining in the water tower water.

If no attempt is made to remove the minerals, alkali, dissolved solids, and salts, you will get high concentrations and these concentrations will leave rock-like deposits on the warmest area of the water system. The inside of the tube sheet and end plates of the water-cooled condenser can be lined with deposits, plugged up, damaged, and the heat exchange process impaired by these high concentrates. Alkali will eat away at tubes and end plates and destroy the condenser. Also, you will have dust, grass, and leaves to deal with as the wind carries these into the open water tower. In time, they will plug the condenser. A good maintenance schedule would call for the condenser to be cleaned once a year.

To clean the condenser, it is necessary to shut down the refrigeration system for a day. The condenser is drained and the end plates are unbolted and removed. On one end, I like to have a temporary ramp set up the length of the condenser, if possible. A barrel is placed at each end of the condenser to catch water runoff. On the opposite end of the condenser from the ramp, there will be a man on a ladder with a three-fourth inch water hose and trigger nozzle. Take a nylon condenser brush and screw it on to a one-eighth inch pipe the length about 5 feet longer than the condenser. The nylon condenser brush has stiff bristles and looks like a bottle brush. The brush is round and the diameter of the brush should be one-eighth to one-fourth inch larger than the inside diameter of the tubes that you are going to clean. The end of the brush has a one-eighth inch pipe nipple that was pressed on by the manufacturer of the brush. The end of the one-eighth inch pipe is placed in the chuck of a one-half inch to three-fourth inch drill motor with a reversing switch. The speed is 300-650 rpm.

The cleaning procedure goes like this: The man with the water hose will pick out one tube and flush it with 2 to 5 gallons water. The man on the ladder and garden hose will keep track of the tubes, the ones that are flushed and scrubbed, and then flushed again, and those that remain to be done.

On the ramp, you will need one man on the drill, one man center way on the pipe to stabilize it, and one man at the tube end plate to see what hole is being flushed and to guide the brush into the tube. After you have the initial flush, the man nearest the tube end plate on the ramp will yell for the water to stop and will place the brush into the tube. The man on the drill will start the drill in forward direction and push the pipe and brush into the tube until the brush shows at the other end.

I always mark the pipe at the drill motor end with tape so I will have proper travel distance and not push the brush beyond the tube plate. The brush could get hung up when the man reverses the drill motor and pulls the pipe and brush out.

After the tube has been brushed forward and backward, the man on the ladder with the garden hose will give another rinse on this tube, with 3 to 5 gallons water. Repeat this same procedure until all the condenser tubes have been brushed clean. Please use a bristle brush. Do not use a steel wire brush. The condenser tubes have a thin wall and you would very much shorten the life of the tubes. Before you attach the brush on to the pipe, put a small rubber cork or wad of rag into the pipe on the end near the brush so that we do not have water traveling on the inside of the pipe into the drill motor chuck and then into the drill motor.

When you have the condenser apart, check the zinc anodes. If they are all eaten up, you will have to replace the zinc anodes. The zinc anode is an 8-inch square plate of zinc, 1¼ inches thick, that will help control galvanic action. Galvanic action is a decay of metal due to the fact that you have more than one metal in the system. Copper tubes and steel shell are the two dissimilar metals that set up a small current and this will cause pitting, decay, and metallic breakdown. The special term is *galvanic action*. Zinc anodes neutralize the little currents and protect the system. If the end plates are cuddy, you might scrape the end plates of the shell and tube condenser with a putty knife to remove mud, slime, and loose scale. Follow the procedures above or use a variation and you will save money and down time.

pH CONTROL

The degree of acidity or alkalinity that exist in the water is called pH. Acidity is the state of being acid or sour. Alkalinity is the state of being high in salts or undissolved solids. Strong alkalies are caustic and will eat up metals in time. Number 7 on the pH scale indicates that the water is neutral and this is an ideal condition for water tower water. Any number less than 7 indicates the water is acid in composition. Any number over 7 indicates that the water is alkali.

If you use untreated water tower condenser water, the water will become high in alkalinity which is caustic and in time the water will eat up the water tower, piping and condenser. You will also have too high a concentration of minerals and undissolved solids in the condenser water and these undissolved solids will deposit themselves in the condenser tubes and end plates.

Water that has too high an acidity will also eat up the water tower, piping, and condenser. Most of the pH testers will cover the range of 6 to 7.6. Determine water pH in this manner:

■ Fill two small vials with 10 milliliters of tower water.

■ Place required drops (3-5) of blue dye (Bromthymol Blue D) into one vial of 10 milliliters of tower water.

■ Place the vials into a Helliage pocket comparator.

■ In the vial that does not have dye, place different color discs until the disc is the same color as the blue dye vial.

■ At this point you can read the pH.

The excessive alkalinity of water tower water with pH beyond 7.2 is lowered and controlled by small amounts of hydrochloric acid or sulpheric acid. The acid will eat the minerals, salts, and alkali and lower pH. The acid is placed in the water tower by an aspirator or small volume precision pump.

When you blow-down a boiler, you pull the lever of your high-pressure valve and keep it open until all the air that is inside of the boiler no longer hisses and gurgles with the water that is leaving the discharge pipe into the drain that is beside the boiler. Also, observe the color of the water. When the water is no longer brown, red, or orange, but comes out clear, you have discharged the harmful undissolved solids, scale, and rust. The blow-down eliminates air pockets which could cause the boiler to air lock and go off the line and you remove harmful undissolved solids. In this process you are also doing a safety check.

Water tower bleed-off is another good aid in pH control. For every gallon of new make-up water to the tower, we try to discharge 18 percent to 25 percent of the concentrated recirculated water. This is done by a flow switch on the water tower supply water which activates a solenoid and gate valve on the pump discharge and the water is discharged into a sewer, a sump, or leach field. A good water discharge could handle 50 percent of your pH program.

If you have a water-cooled condenser where the supply is 100 percent ocean, lake, or stream, with *no* recirculation, you do not need pH control. Just pump the water from the stream, lake, or ocean into the condenser and return to the same. You do not need pH control in a closed system because no new water is added and there is no evaporation leaving large amounts of undissolved solids.

The water that is in the pan of the closed system water tower that is pumped to the top of the tower and splotches over the tube nest or bundle and then falls into the pan, must have water chemistry. If the water is too cold (below 75 degrees F.), you will have to throttle the water to maintain sufficient head pressure. Too low a head pressure will cause the refrigeration oil to leave the compressor crankcase and migrate.

When an individual changes carboys and works with these acids, he must wear rubber gloves and safety glasses. If you should get acid on you, quickly put your skin, with the acid on it, into the water of the water tower. Water will neutralize hydrochloric and sulphuric acid. Extreme care should be used with these acids. There is no need to spill it, be sloppy, or get it on you or your clothes. If you get acid on your clothes, quickly remove the garment and neutralize the acid by immersing the garment in the water tower water. I have gotten hydrochloric acid on my hands. This acid is slow to work and is not hot. I just washed my hands in the water tower water. Sulphuric acid is very hot and fast acting. The first rubber gloves I used were too thin and I could feel the heat of the sulphuric acid even when I immersed my hands and gloves in water. Use heavy rubber gloves. If you have to work with sulphuric acid, such as moving it, running hoses, installing acid pumps, etc., you should wear a rubber rain hat, raincoat, rubber gloves, galoshes, and safety glasses. Remember, sulphuric acid will burn and blister your skin before you can get the acid garments off and the skin into water. Most importantly, safety glasses are a must when you are working with acids.

BIOCIDES

Biocides are chemicals that are used in small amounts to kill bacteria. Chlorine that is used in a gas form is an excellent biocide. On the average water tower, I like to use 1½ pounds of pure chlorine gas added to the recirculating water by an aspirator that is on the discharge of the water tower pump. Take one hour of time to add the gas and do this process every 24 hours. This process is automatic by a time clock and you can keep track of the pounds of gas by placing your gas cylinder and regulator on a scale. Note the difference of weight that occurs every 24 hours. Water tower water, when the conditions and temperature are right, will have vermin, fungus, or become like a cesspool. On the market, you will find many good biocides. Pick a good biocide from a reputable company and use it on a regular schedule. I like chlorine gas because it is a herbicide as well as a biocide. When you use chlorine gas, you "kill two birds with one stone."

DEFOAM SOLUTIONS

When you are using the biocide, you will find that the dead bacteria, combined with the minerals in the water, will form soap suds foam on the water surface in the bottom pans of the water tower. When the fans come on, soap suds foam is blown all over for a quarter of a mile radius. It makes splotches on windows, car paint, washed clothing, and is a plain nuisance. Defoam is an oil-base solution that you add to the bottom pan of the water tower and its quick action will distill the soap suds foam. Use two coffee cups of defoam and dribble it over the foam that is in the pans of the water tower. This should rid the water tower of foam in five minutes. Try to keep a gallon of defoam on hand. This will last a long time. After you treat the tower water with biocide, recheck it after 40 minutes for foam.

HERBICIDES

A herbicide is a chemical solution that is added to water tower water to kill plant growth, moss, and slime. Buy the herbicide from a reputable dealer and set up a schedule and stick to it. Chlorine gas will do the herbicide as well as the biocide. I have seen some water towers filled with moss or slime and you could have grown a crop of corn in the water tower if you wanted to. The moss and slime could plug the condenser or "y" screens that are before the condenser. A lot of labor and down time is connected with this problem.

PHOSPHATE BALLS

Many firms will add one phosphate ball to the water tower water daily. Normally, the ball is dissolved in a tank and pumped in with a small volume pump that is brought on the line by the same flow switch that triggers the water tower discharge solenoid valve. The purpose of the phosphate is to keep dissolved solids in solution and prevent the precipitation of the dissolved solids in the condenser.

CHROMATE PELLETS (SODIUM)

These are used in closed chill water systems and their purpose is to prevent corrosion by laying a coating of chromate ion on the inside surface of the chill water system. Chromate pellets are a poison and cannot be used near potable water. You should maintain 700 to 1000 parts per million of chromate in your chill water system and you check this by use of a Taylor comparator. The chromate will not hurt the pump seals, but care must be taken in putting the chromate into the chill water system. Chromate will foam up when added to the chill water and is a poison. You will have to wash your hands well after application.

LOW-PRESSURE BOILERS

With low-pressure boilers, try to maintain a level of sodium silicate of 80-100 parts per million. The purpose of sodium silicate is to prevent corrosion inside the heating system. Once a week with low-pressure boilers, blow down the boiler using the McDonald Miller high-pressure, pop-off valve.

When you blow down a boiler, pull the lever of the high-pressure valve and keep it open until all the air that is inside of the boiler no longer hisses and gurgles with the water that is leaving the discharge pipe into the drain that is beside the boiler. Also, observe the color of the water. When the color is no longer brown, red, or orange, but comes out clear, you have discharged the harmful undissolved solids, scale, and rust. The blow-down eliminates air pockets which could cause the boiler to air lock and go off the line and we remove harmful undissolved solids. In this process you are also doing a safety check to see if the high pressure pop-off valve is in working order. Boilers have an automatic make up by way of a water pressure regulator valve. You will come up to proper sight glass level automatically with no worry of the boiler being low on water.

The cooling tower's main function is to exchange heat and to save water by recycling. The water can be recycled to the point of 4½ times the total of dissolved solids of the raw supply of water. If you have a 25 percent discharge on the water tower, you will have close to a 4 to 1 concentration. This is done with a water meter on the discharge and a water meter on the raw supply tower water. Adjust the discharge gate valve to 25 percent in gallons of the raw supply water. This will equal the 4 to 1 concentration. A dissolved solids meter is helpful in reading the required relationship.

Example: The dissolved solids meter reads the raw supply water at 265 parts per million (PPM). Multiply 265 PPM times 4.5 and that will gives you a figure of 1,192 PPM. That is the maximum concentration that you should have for the recirculating water tower water. The next step is take the dissolved solid meter and read a sample of the water tower water that you have obtained from the pan of the water tower. If the concentrations are too high, you will increase the discharge. If the concentrations are too low, you can save water and shut down the discharge a slight bit so there is less water being discharged.

Note: The Environmental Protection Agency may limit the use of certain chemicals in water-treatment programs.

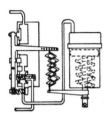

Chapter 17
Tools & Materials

A vacuum pump, single- or two-stage, that is motor direct coupled should have a handle on it and one man must be able to carry the vacuum pump on roofs, in attics, or small confined places with no trouble. It should weigh no more than 50 pounds. A one-quarter or one-half horsepower motor is good enough.

A welding torch or welding unit must be a small portable unit with a rack that has a handle on it which holds the bottles of gas. The welding unit can be one- or two-stage, must weigh less than 50 pounds and can be carried by one man.

A small A, B, or C fire extinguisher should be part of your equipment in case of insulation, refrigeration oil, electrical, wood flammable liquid fire. Do not weld without a fire extinguisher.

Use safety glasses on silver brazing, working with refrigerants, working around rotating type machines, changing filters, working on ducts when the air is flowing through, and working with acids. You will find that welding goggles are too dark for silver brazing. However, use welding goggles on all other types of welding.

I like a battery-operated leak checker that makes a sound or flashes a light to indicate a small leak. Stay away from the 120-volt leak detectors as many times you will not have voltage nearby. If we are dealing with leaks that are quite noticeable, a halide leak detector or a bottle that is one-third soap and two-thirds water solution with a small brush or applicator will do. Do not buy a charging cylinder unless your work is over 50 percent critical charge units.

Tools to work on copper tubing include a cutter with reamer attached, clamp off, flare tools, swedge tools, a hacksaw, fine and course files, a sheet of fine, medium, and course emery paper, hand tools with tool box, a rachet set, and a ball pean hammer.

Use a small size volt, ohm, ammeter. Voltage will be 0 to 500 volts ac. Use a low voltage range to accurately read 24 volts ac. You must be able to read very clearly 1 to 30 ohms. The average low cost meter will handle the upper ohms, low current dc amps and millivolts for thermocouples.

A clamp-on ammeter must be able to go as high as 300 amps ac. A Wiggington voltage checker, 120, 240, 480 volts ac, made by Square D is rugged and dependable. You should have refrigeration gauges and charging hoses attached. The length of hoses should be 36 inches.

Other items you should have are a three-eighth inch drill motor and drill bits. A drop light with 50 feet of cord. A flashlight. A grease gun with hose, 90-degree zerk and cartridges of grease. You must have the hose and 90-degree zerk to get many bearings. The following is a list of materials you will need: Tube of still floss silver solder sticks to silver braze copper to copper (melts at approximately 1800 degrees F); small roll of easy flow 45 silver solder (45 percent silver content) to silver braze copper to steel or copper to brass—melts at approximately 1400 degrees F; small tin of Borax flux to clean work and help silver solder flow; a few one-fourth inch flare nuts; vacuum pump oil; refrigeration oil; Schraeder valve stems (copper); and piercing valves (3 sizes) to charge systems that do not have service entrance valves.

In addition have some refrigerant, R 12 and R 22 in throwaway drums, size 25- or 30-pounders. If you are working large tonnage, you will be using the 125-pound cylinders or 250-pound drums of R 11. When working units under 10 tons, use the 25- to 30-pound throwaway bottles.

You will also need jumper wires with alligator clips. Have five of these jumpers, a roll of electrical tape and a roll of duct tape.

MAINTENANCE TIPS

In Fig. 17-1, you see the fan shaft with a pillow block bearing at the A lube fitting. There is a pillow block bearing at the B lube fitting. This is a typical fan shaft and bearing setup. For usual running time, I would work the lever on the grease gun three times every six months. When the fan is on the line 24 hours a day, I would work the lever on the gun three times every three months.

In Fig. 17-2, you see a typical oil cup bearing. The little lid on the cup is spring loaded. For usual fan running, just fill the cup every six months. If the fan is run 24 hours a day, fill the cup up every three months. Use 10- to 30-weight oil.

BELTS & PULLEYS

You should have belts on hand to maintain the fans that are on continuous duty. Different manufacturers will use different numbers for the same size belt. Belts are sized by the numbers two ways: by width and

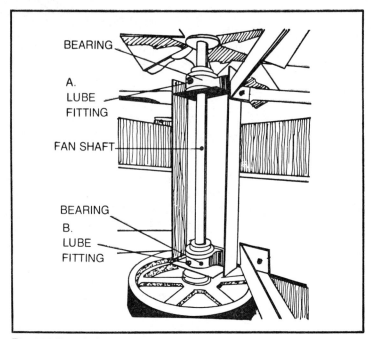

Fig. 17-1. Fan shaft.

by perimeter. Perimeter is belt length or distance measured around the pulleys. See Fig. 17-3.

When I do not know what the belt size is, I take a string and hold it on point 1 of pulley A (Fig. 17-3) and extend the string to point 2 of pulley B and wrap the string around pulley B to point 3. Then continue from 3 to point 4 and wrap around pulley A to point 1 and cut the string. Next, take the string off the pulleys and measure the length with a ruler. If the string is 38 inches long, this is the perimeter. Measure the width of the pulley grove or belt that is worn out. Using this information in Figs. 17-4 and 17-5, you can find the belt sizes. Use this method when the belt wears out and you cannot read the stamped or printed code number.

Automotive belts have a special system. Specify air conditioning belt because the air conditioning belt has steel wire in it—not fiber. There will be a green line that runs around the air conditioning belt as a marker. When a car is going 65 miles per hour, the compressor is capable of developing 5 tons refrigeration. Tonnage is directly proportional to rpm and an air conditioning belt is needed because it is heavy duty, more durable and will hold up to hard usage.

A belt with the size numbers 4L 380, or 2380, or A 38 is the *same belt*. When you use Table 17-1 chart No. 3, the number of inches is written in with no fraction. Chart No. 3 belts are industrial belts with more filler in the belt. These belts are more rugged and heavy duty. In Table 17-1 chart

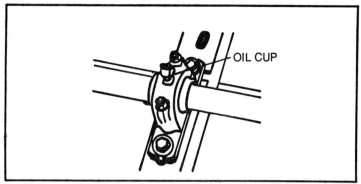

Fig. 17-2. A typical oil cup bearing.

No. 2, the manufacturer will use the number for the width ahead of the perimeter. The perimeter will be three digit and the last digit will be a fraction in tenths. All belts are printed or stamped with this code.

The letters VP printed on a belt means *variable pitch* in reference to drive pully. The sheaves can be adjusted to vary it's diameter. The sides of a pully are known as shives. Every pully has two shives. If the letter X appears in the belt number, you have a notched belt. A notched belt is used on small pulleys where a standard size belt would slip because the pulley has too much bend for the belt. A notched belt will bend a great deal more than a standard V belt. Multiple belts on a fan are ordered as a matched set. You must have matched belts with exact perimeter or one belt will be too tight, one too loose, one will flop, and one might come off. You need a matched set to adjust belts for true running and proper tension on multiple belts.

When you remove air conditioning belts from automotive air conditioning units, you unscrew the bolt that holds the idler pulley tense and remove the tension from the belts. Then just slip the belts off the pulleys. Sometimes you will have to remove other belts by undoing the tension and slipping them off the pulleys. To remove other fan belts, first shut off the unit and remove the belt guard. Next, you cock or wedge the belt to the side of the largest pulley and roll and wedge it off. It is not hard

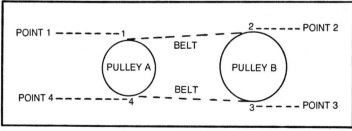

Fig. 17-3. Perimeter of a pulley.

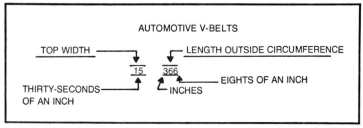

Fig. 17-4. The first two numbers is the top width of the automotive V-belt measured in thirty-seconds of an inch. For example, this belt size indicates a width of fifteen-thirty-seconds of an inch. The next three numbers (366) tell you the perimeter of the belt. The number 36 tells you the belt is 36 inches in perimeter. Cut the belt and the length will be 36 inches. The last number 6 of the 366 tells you that the belt is an additional six-eighths of an inch long. The total belt length (perimeter) is 36 6/8 inches. The belt is fifteen-thirty-seconds of an inch wide and 36 6/8 inches long.

to do and works fast. Sometimes I will take a knife and cut the belts off. In a laboratory where I work, I will change a belt trio without pulling the belt guard and have the unit down less than five minutes. The fans are on the line 24 hours a day and not to be shut down.

When putting belts on, you wrap the belt over the motor pulley which is normally the small pulley and is vari-pitch. Then you place the belt on top of the large pulley and turn the pulley until all the slack is gone. Use a little more muscle and turn the large pulley and the belt will snap on the pulley. If it was adjusted properly on the worn out belt, the new belt will stretch over the large pulley with no damage. You should be able to push the new belt in one-fourth to one-half inch. If the belt is flopping with vibration noise, tighten until the flop goes away and the belt runs quiet.

When belts squeek or squeal, try this little trick. Roll up your shirt sleeve so the sleeve does not get caught in the running belts. Take a dry bar of hand soap and hold it on each side of the V-belt while the fan is running. The glycerin in the soap is a good belt dressing and will kill the noise.

When you check the belts, turn off the motor and stop the unit. With a flashlight, slowly turn the belts and look for cracks, tears, scorch and excessive wear. If you have multiple belts, look for loose, stretched or

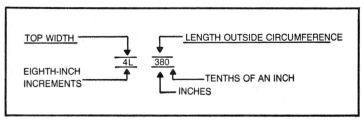

Fig. 17-5. Fractional horsepower V-belts (non-automotive).

Table 17-1. Width Charts.

WIDTH CHART #1	WIDTH CHART #2	WIDTH CHART #3
2L---¼ inch	0---¼ inch	no #
3L---⅜ inch	1---⅜ inch	no #
4L---½ inch	2---½ inch	A---½ inch
5L---⅝ inch	3---⅝ inch	B---⅝ inch
		C---¾ inch

floppy belts in the belt set. Either adjust the motor to remove the flop, replace the stretched belt, or install a new matched belt set.

Most belts are neopreme and their recommended shelf life is four years. After your years, the weather will dry them where they will crack and lose their flexibility. A neopreme belt will not stretch under heat and load conditions. Shelf life can be extended on belts by making the belt a coil, sealing the coiled belt in plastic, and placing it on a dry shelf.

Belt tension can be adjusted on V-belts by raising or lowering the motor mounts or opening or closing the variable pitch pulley that is on the motor. The fan speed will change a slight bit when you adjust the variable speed pulley. The length of the belt does not effect fan speed. The speed remains the same whether the belt is 2 feet or 12 in perimeter. Only pulleys and their different sizes change belt speed. See Fig. 17-6.

When the driven pulley stays the same diameter and the driver pulley remains at 1800 rpm, then the speed of the driven is directly proportional to the diameter of the driver. See Table 17-2. When you increase fan rpm, use your clamp-on ammeter and do not exceed motor current. This will be the motor name plate current times the service factor. If there is no service factor, then use name plate motor currents. If the current is too high, lower fan rpm which will lower the motor current.

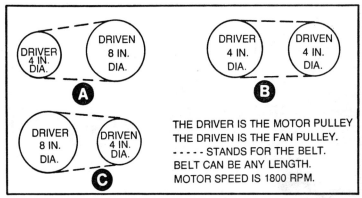

Fig. 17-6. Belt tension.

Table 17-2. Pulley Diameter and Speed.

DIAMETER OF DRIVER PULLEY	DRIVEN PULLEY SPEED
4 inches	900 rpm
6 inches	1080 rpm
8 inches	1800 rpm
10 inches	2333 rpm
12 inches	2700 rpm

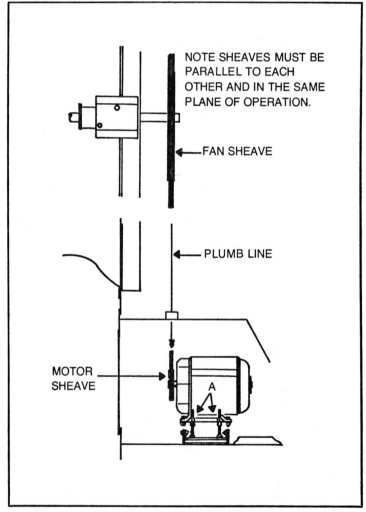

NOTE SHEAVES MUST BE PARALLEL TO EACH OTHER AND IN THE SAME PLANE OF OPERATION.

FAN SHEAVE

PLUMB LINE

MOTOR SHEAVE

A

Fig. 17-7. Sheave alignment.

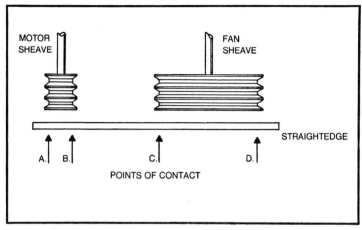

Fig. 17-8. Checking sheave alignment.

ALIGNMENT OF SHEAVES OR PULLEYS

Whenever you replace motors or pulleys, one problem that must be solved is pulley alignment. A sheave or pulley is the same thing. After you have installed the new motor or sheaves, there are two ways to approach alignment. When the two sheaves are directly over each other, you can use a plumb line, as shown in Fig. 17-7. You can move either the motor or fan sheave on the shaft until the plumb line is dead center to the motor sheave. Then put your belt around the smaller sheave and roll it over the larger sheave. If you cannot roll the belt over the larger sheave, give yourself some slack by loosening the motor adjusting nuts at point A.

On a multi-belt pulley, you must make sure the individual V belt openings are the same for all belts. Then pick one V-belt V and extend the plumb line to the corresponding V on the pulley below.

Figure 17-8 shows sheave alignment by using a straight edge. You can move either the motor sheave, fan sheave, or both motor and fan sheaves on their respective shafts until the straight edge makes contact at points A, B, C, and D. The straight edge can be used horizontally and vertically where a plumb line could only be used vertically.

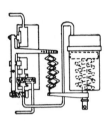

Chapter 18
Computer Air Conditioning

The purpose of computerized air conditioning is to monitor the temperature of rooms, laboratories, process cooling waters, and chambers, to turn on and to turn off fans, pumps, motors, compressors, to give alarms, to show schematic read outs of fans, piping and rooms—plus provide management with control over energy and temperature. The fire alarms and security are integrated in most computer systems for air conditioning. To do the above functions the hardware needed is a computer with a program unit (which is a fast turning tape or disc), a teletypewriter printer, a CRT terminal (which is a small television monitor with key board), wiring, controls, printed circuit card panels, and electric/pneumatic panels.

The electric/pneumatic panels are to transition the air controls of the refrigeration, air conditioning and heating systems to the computer. The teletypewriter printer is a machine that prints alarms, changes of state or mode (when a compressor comes on the line this printer prints the change of state), alarm acknowledgements, commands, and program changes.

When a programmer or supervisor changes the program, he will use the typewriter of the teletypewriter printer to make the required change. The management will read the printout sheet of the printer to control and be aware of the plant refrigeration, air conditioning, heating, security and fire alarms.

Suppose you are the air conditioning man and you walk into the computer room. The first thing you will do is to go to the printer and see what trouble calls or alarms have been printed and the time and date. Every time a motor, fan, pump or compressor comes on, the printer will record it on paper. When the temperatures exceed guidelines, the printer

140

will record it as well as time and date. This is called an alarm. All the acknowledged and current alarms will appear on a CRT display when you push the alarm summary key.

CRT TERMINAL

There are four important keys that you will use as operator on a typical CRT terminal. When you push the *clear key*, whatever is on the screen will disappear and instructions will come on the screen. The little TV screen on the CRT terminal is also called a display. For example, something like below will appear:

IDLE PAGE

Press --- Alarm Sum key --- for alarm summary

Press --- Index --- for building/ system/ point selection

Press --- Inform --- For general information

When you press the alarm summary key, a display will appear on the CRT terminal of acknowledged and current alarms.

When you push the *index key*, this gives you a breakdown of buildings, floors, different systems or operations. For example, if you had five buildings, it would read:

BUILDING NAME

Building 100
Building 200
Building 300
Building 400
Building 500

If you wanted to find out what Building 100 is composed of, you would type the number 100 on the keyboard of the CRT terminal and push the return key. You would get a readout and at the bottom of the readout you will read "continue on next page - to proceed, enter p (Return)." This means if our information is not on this display we push the typewriter key, letter p, and push the return key. This will permit you to go to the next display page. A computer is like a book with pages. To go from one page to another you hit the letter p key and the return key. On the display, you will get a readout of fans, pumps, out water systems, boilers, or whatever information is catalogued.

Suppose you decide to look at fan number AS4U that is in Building 100. The building for this fan and the fan appears in the readout of the display of the CRT terminal. Punch in the keys AS4U and the return key on the CRT terminal. You will find the fan with a list of numbered items will appear like the example below.

FAN AS4U

Point   1.   Motor control

------ On

	Point	2.	Cold plenum temperature 47 (62) 72
	Point	3.	Set Point -----5.2° F.
	Point	4.	Hot Plenum- Left side temp. 80(95) 105
	Point	5.	Hot Plenum - Set Point --+25° F.
CRI	Point	6.	Hot Plenum - Right side Temp. 81 (73) 106
	Point	7.	Hot Plenum - Set Point --+25° F.
	Point	8.	Motor control mode --- Automatic

Suppose the "hot plenum - right side temp." is the reading. What the CRT display is telling you is that the programmer has decided to carry a temperature between 81 degrees F to 106 degrees F. The number 73 that appears in the ( ) is the actual temperature of the plenum. In this case 73 degrees F is not between 81 and 106 degrees F. Point 6 is showing an alarm condition. CRI stands for *critical alarm*. The entire line of Point 6 would be flashing on and off to call your attention. The teletypewriter printer has typed the same information plus the time and date. I will cover the acknowledgement of an alarm ladder on in this chapter.

The information key might have information on the fans such as maintenance performed, greased, oiled, belts and belt numbers, and motor information. Some computers will give schematics on hot and cold decks, hot water, chill water, glycol water, and process cooling water systems. A schematic on fan ducts is important because it will show the hot and cold decks with all the dampers and zones.

When you have a frozen damper or air motor with leaky air diaphram, the information on the schematic will locate the damper and zone—for example, the fifth zone on the right side of the fan. When you are in the attic or dropped ceiling you must know where the zone and damper is by schematic. There are ducts, pipes, conduits and insulation everywhere and there is difficulty in locating the damper and zone.

When you operate a computer, one of the first things you learn is that there are different access levels. These levels are for the purpose of tight control over the use of energy, to keep the cost of heating and cooling as low as possible, prohibit unauthorized people from tampering with heating and air conditioning, give more control over the use of heating, air conditioning, and refrigeration, and save labor costs.

SAVING ENERGY

If the outside weather is 65 degrees F or less, you can have the computer programmed to raise the chill water and cold deck temperature. The cooling requirements will be minimum. There is no need for a 40-degree F chill water and a 50-degree F cold deck. The computer automatically puts the chill water to 50 degrees F and the cold deck to 60 degrees F. The boilers will automatically throttle back and use less fuel because some of the boiler heat was consumed by the chill water and cold deck. The compressors will use less electricity because you need less horsepower to maintain higher chill water temperatures. The savings is in electricity, oil, or natural gas.

When the outside weather is plus 90 degrees, the computer can be programmed to carry the chill water 40 to 47 degrees F and lowers the hot deck temperature to 85 degrees F. By using lower hot deck temperature, the compressor load will decrease as some of the load is consumed by the boilers as well as the building. The savings is in electricity, oil, or natural gas. The computer will give more control by turning on and off pumps, boilers, air conditioning, compressors, water towers and fans automatically. The computer can run the air conditioning over the weekends, during the evenings and save the wages of a man being on duty or standby.

There are four levels of operation of a computer. There is a guard level, operator level, supervisor level, and programmer level.

OPERATOR LEVEL

The *operator level* will let you do what the *guard level* can do plus more. It can acknowledge alarms, turn on and off fans, motors, pumps, put fans on manual control or automatic control, plus have the computer give us a read out at different time intervals on the temperatures of rooms, chill waters, decks or process cooling waters and shut itself off automatically.

SUPERVISOR LEVEL

The *supervisor level* can do what the guard and operator level can do, plus more. The supervisor level can change the temperatures (this is done by changing the set points), alarm limits, program times plus add or delete set points on the computer program. The supervisor can remove your identification and password.

PROGRAMMER LEVEL

The programmer level is the highest level on the computer. He can do guard, operator, supervisor plus change internal computer program. The computer programmer probably is the manufacturer or his representative who puts the program on the computer using the teletypewriter unit. He will make all major changes. Minor changes are done at the supervisor level.

LOG ON THE COMPUTER

To log on the computer, use the CRT unit as follows:
Push the CLEAR key and bring on the IDLE page.
Type the word hello.
The CRT unit will display the word "hello."
Type your initials and push the space bar.
The CRT unit will display the word password.

Type the password and the return key. Most passwords are three letters. My password is run. I type in the word run and hit the return key. My initials appear on the top right corner of the CRT display and the teletypewriter printer unit prints R.A.F. logged in at time and date. When you type the password it does not appear on the CRT display. To acknowledge an alarm, first log in by the above procedure. Press the alarm summary key. This will bring on the alarm: CRI "Bld. 100, AS4U, Point 6." plus other alarms you may have on the system.

Type "100, AS4U, Point 6" and you get the display. Point 6 will be flashing on and off to get your attention. Push the *alarm acknowledgement* key. The teletypewriter printer prints Fan alarm acknowledged, R.A.F., date and time.

Push general information key and get a schematic on fan AS4U. You have to find out where the zone, damper, and 3 way hot water valve is for this fan. Suppose the schematic shows the zone to be the fifth one on the right of fan AS4U. Let's go into the attic and take a look and see.

Why is the hot plenum on the right side too cold? You find that the three-way heating valve is frozen closed. With tools and high penetrating oil, work the plunger up and down and the three-way valve starts working. The hot deck temperature comes up to 90 degrees F and the teleprinter prints "Alarm cleared, Build. 100, Fan AS4U, Point 6." and the time and date. The CRT display no longer shows AS4U in the alarm summary.

When you leave the computer room, log off using the CRT unit. Push the CLEAR key and bring on the IDLE PAGE. Type the word "bye" and push the return key. The CRT display will read PASSWORD. Type in the password for your initials. Your initials will disappear from the right corner of the CRT unit display and the teletypewriter printer will print your initials, logged off, date and time. You must log off, otherwise other people will be able to give commands and you will be responsible.

Turn on a fan using the computer. This operation can be done with the CRT unit. The same procedure will work for all commands on pumps, temperatures (rooms, decks, and waters), compressor boilers.

■ Log on the computer.

■ If you don't know the building, push the information key and break down the plant until you find the building and fan. Suppose you want to turn on fan AS4U. You have found that this fan is in building 100.

■ Push the INDEX key and the key on the CRT keyboard. The CRT display will read:

Building 100
Building 200
Building 300
Building 400
Building 500

■ Push 100 and the return key. The CRT will show all fans, pumps, boilers, heating and air conditioning equipment for building 100.

■ When fan AS4U comes on the display, type on the CRT terminal AS4U and push the return key. This brings on the display.

■ We will work with point 1. Because the fan is off the motor control would say off. Imagine that point 1 says:

Point   1.   Motor control - - - - off

■ Push the number 1 and the return key on the CRT terminal. This whole operation is done with the CRT terminal. Point 1 will flash on and off and you now know that you have raised point 1. You must raise the point that you want to command.

■ There is a START key and a STOP key on the CRT terminal. Push the START key and the return key. The teletypewriter printer will print FAN TURNED ON AS4U, date and time. You know who turned the fan on, because the log sheet or the teletypewriter printer sheet tells you who is logged on.

Point 1 on display of the CRT unit will read like this:

Point   1.   Motor control - - - - on

You can turn the fan off with this same procedure. Just push 1 and the return key. This will raise the point. Then push the stop key and the return key and the fan will turn off. You will use this procedure on any command that you wish to initiate. Remember you are going through the pages of the computer. Other systems will be close to this one in operation.

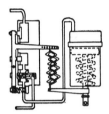

Chapter 19
Fan Systems

In the air conditioning, refrigeration and heating field there are two primary types of fans: the propeller fans and the squirrel cage fans. The propeller fan will be three or more blades at a 20- to 45-degree pitch. The wheel of the squirrel cage fan is divided into three catagories: air foil backward curve, flat blade backward curve, and forward curve.

FAN DATA

A propeller fan has a shroud that will fit around the propeller and its purpose is to direct and concentrate the air flow. An example is an automative air conditioning. Propeller fans are sized by diameter and pitch and number of blades.

The squirrel cage wheel has a scroll or envelope that fits around the wheel and its purpose is to direct and concentrate the air flow. Squirrel cage fans are sized by wheel size, cubic feet per minute of air flow (cfm), and the air foil backward curve blade, flat backward blade or forward curve blade. The air foil blade (Fig. 19-1) looks like the wing of a toy airplane. It will deliver 5 to 7 % more cfm per horse power over a same size backward curve flat blade. When the backward curve wheel turns, the blades of the wheel are swatting the air. Figure 19-2(A) shows a single-width, single-inlet, backward-curve flat-blade, squirrel-cage wheel. The arrow shows the direction of rotation of the wheel. Figure 19-2(B) shows a double-width, double-inlet, backward-curve, flat-blade, squirrel cage wheel. The abbreviation for single-width, single-inlet wheel is SWSI. The abbreviation for double-width, double-inlet wheel is DWDI.

On the DWDI wheel, the fan maker will many times use two SWSI wheels and mount them back to back with nut and bolt, rivets or spot

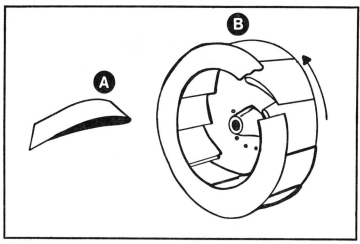

Fig. 19-1. A cross section of an air foil blade (A). There are many of these blades in a wheel. An air foil backward curve squirrel cage wheel (B) is single-width. The arrow shows the direction of rotation.

welding. A lot of time has been lost on the job because the wheel was improperly installed in the scroll.

Follow my arrows because rotation is critical from the standpoint of placing the wheel in the scroll and wiring the motor to turn the wheel the right direction. A burnt out fan motor or a job that is done by more than one man brings this problem into clear focus.

A fan maker will show a picture of the wheel in the scroll and place the abbreviation SWSI or DWDI under the picture to describe the fan. The wheel and scroll are considered as one unit in the trade. SWSI or DWDI as also known as the style of fan.

Figure 19-3 is a single-width, single-inlet, forward-curve, squirrel-cage fan. This fan also comes in double-width, double-inlet. The blade of this wheel is cupping the air and throwing the air forward.

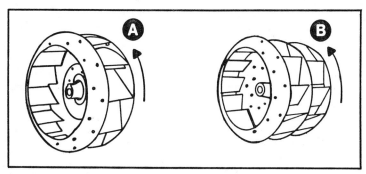

Fig. 19-2. Fan blades.

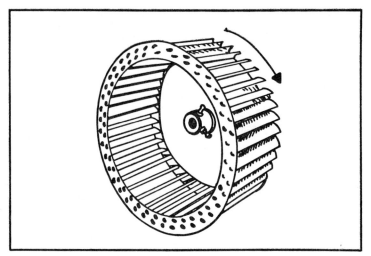

Fig. 19-3. A squirrel-cage fan.

Figure 19-4 is a single-width, single-inlet fan. There is one wheel and it is a backward curve. The round opening is suction and the rectangular opening is discharge. These fans are weather proof and may be outside the building, inside the building, in attics or basements. These fans are exhaust fans. However, if the fan scroll was insulated, it could be used for air conditioning.

Figure 19-5 is a fan that is double-width, double-inlet. This fan has two wheels that are backward curve flat blade. The air comes in on both sides of the fan at the side where the motor is and at the side where you see the pillow block bearing. This fan has two suctions and the rectangular opening where the double wheel can be seen is the common discharge. The fan is direct drive and the motor shaft must be in perfect alignment with the fan shaft. The motor and fan shafts are joined by a flange coupling. The fan is mounted on rails with springs as vibration insulators. The motor is a drip proof motor. There are air vent openings on the bottom of both end bells. This is a pedestal mount motor with permanent seal bearings.

Figure 19-6 shows two views of the same fan. This is a double-width, double-inlet with the wheels mounted back-to-back. Note the pillow block bearings and they have grease zerk fittings on them. Fans like this are used for air conditioning and they are on the line 24 hours a day. Grease the pillow block bearings every three months. This is an industrial rated fan, heavy duty. There are screens on the suction on both sides to catch airborne matter.

Figure 19-7(A) shows a motor driven pulley and a fan driven pulley. There is a matched trio belt set on the pulleys. Figure 19-7(B) shows the belt guard covering the pulleys and belts. Always replace the belt guard when you install new belts. A fan that does not have the belt guard on is a

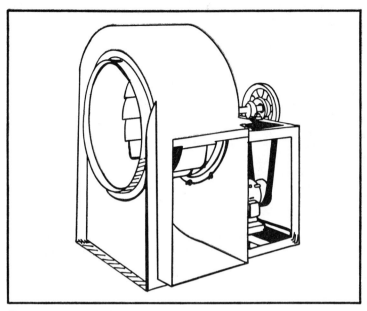

Fig. 19-4. A single-width, single-inlet fan.

safety hazard and you can be cited and fined by the Occupational Safety and Health Administration.

RULES FOR PROPELLER FAN ROTATION

The propeller fan must turn in the direction so that the force exerted on the fan shaft is applied directly against the thrust bearing. The thrust

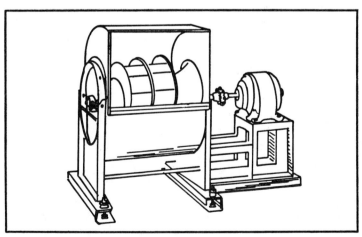

Fig. 19-5. A double-width, double-inlet fan.

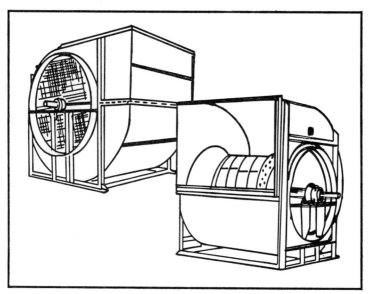

Fig. 19-6. A double-width, double-inlet fan with wheels mounted back-to-back.

bearing can be on the end of a motor or on a pulley shaft. When you turn the propeller in one direction, you will move a volume of air. If you reverse rotation, the volume of air will diminish and you will feel like you are churning the air. When you are moving a volume of air, you are turning the propeller in the right direction. Most important is to install the propeller on the fan shaft so that the force the fan developes on the shaft in turning is directed against the thrust washer or bearing. Many times the propeller is like the forward curve blade. We cup the air and throw it forward. Rotation is critical.

Figure 19-8 shows a propeller fan. The square sheet metal enclosure around the fan blade is called a shroud. Note the arrows that show the

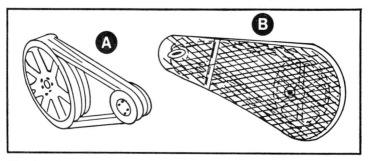

Fig. 19-7. A motor-driven pulley and a fan-driven pulley (A), and a belt guard (B).

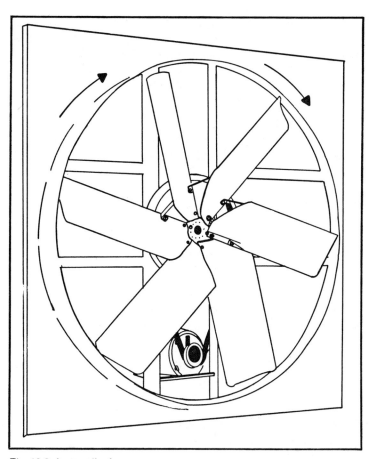

Fig. 19-8. A propeller fan.

rotation of the fan. This particular fan cups the air and throws the air forward. This is a heavy duty industrial rated fan. The shroud and all the fan metal is heavy plate with steel structural reinforcement. Even the blades are supported by steel brackets. The fan blades are mounted on the driven pulley and the pulley and blades are one integral part. The pulley will give this fan rigidity and strength.

Fans that look like Fig. 19-9 slice the air and cast the air forward. There is a guard on the back that is mounted on the fan shroud. The arrows show rotation.

The propeller fan that is equal in diameter to the forward curve squirrel cage fan will deliver less cfm air per same rpm. It makes no difference the number of blades or pitch.

With the squirrel cage fan, we must move the air in the direction of the discharge of the scroll or envelope. Many large fans have a cabinet that will have two or more scrolls. The scrolls will be DWDI with double wheels. All the wheels are mounted on the same fan shaft and the shaft may be two pieces joined by a flange. The fan shaft will be supported by pillow block bearings, flange or rigid mount. Fan shafts on large fans may be hollow or solid. The fan shaft may be cut down (reduced in diameter) at the pillow block bearings. The fan shaft diameter is important because of bearing ID, pulley ID, fan wheel or propeller ID, and coupling ID if it is a direct-driven fan.

The tip speed of a propeller or squirrel cage fan is where you have reached the fans maximum capacity to move air and you are starting to churn the air.

INCREASING THE RPM OF FANS

■ Do not exceed the service factor of the motor current that is printed on the name plate of the motor. Keep your eye on the clamp-on ampere meter when the fan is running.

■ Keep below tip speed. When the air flow starts to diminish and you start to churn the air, you are at too high a speed. Drop the speed until you get good movement of air with no churning.

■ Listen for vibration and excess noise. When you go beyond the design speed of bearings or shafts, they will vibrate and set up harmonic noise.

■ Make sure of good belt tension.

■ The speed range of the forward curve wheels, 1 to 1500 rpm and 1 to 2800 rpm is the speed range of the backward curve wheels. You might go as high as 3200 rpm on backward curve wheels on some fans.

■ The current on forward curve fan motors will rise quickly with small increases of fan rpm. The forward curve fan with the same wheel diameter as the backward curve fan will move much more air per the same rpm. The backward curve fan will have to run much faster to move the same cfm of air as the forward curve fan (Wheels being the same diameter). The advantage of the backward curve wheel over the forward curve wheel is that you can crank up the rpm a great deal with out such high rise of motor current.

When there is additional heat load, more exhaust needed, or more zones added to a fan system is when you increase the fan rpm which will increase cfm. When you increase fan rpm, follow the preceeding steps. With the increase fan rpm make sure the fixed dampers and zone dampers are open.

THROTTLING OF FANS

Fans can be throttled with a fixed damper or gate placed in the return air (suction) of the fan system. *Do not* use the fixed damper or gate on the discharge. The stress can blow the ducts apart. When you throttle the cfm down, the motor current will fall. Less air moved (cfm) means less motor

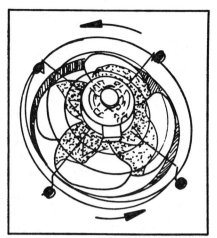

Fig. 19-9. This type of fan casts
the air forward.

current. The fan may also be throttled by using the dampers in the room
registers. You will become involved with throttling when you do air
balance work. The air that is supplied to a room should approximately
equal the air that leaves the room. An exception is clean rooms. The
separate return air of a clean room is closed a slight amount in order to get
a manometer reading of one-tenth inch water column plus pressure inside
the clean room in reference to the air outside the clean room walls and
ceiling. This is to keep the dirt from migrating into the clean room cracks
and open space under doors.

Centrifical water pumps are throttled on the discharge side only.
Never throttle a water pump on suction. If the suction is diminished, the
water pump will cavatate, loose prime, or burn out it's seals. To throttle a
water pump, you turn the gate valve on the discharge. Throttle water
pumps where motor currents are too high or reduced flow of water is
required. I have seen some pumps that were throttled when they were
installed new to keep the motors from burning up. The discharge gate
valve was turned closed until the motor current dropped below the service
factor on the motor name plate. The discharge gate valve was tied off and
tagged. A good air condition and heating man might have to throttle pumps
as well as fans. A dirty air filter will throttle your fan and reduce motor
current.

FAN VIBRATION

■ Check to see if you have dirt build up on the fan blades which has
upset the balance of the wheel. Scrape the fan blade clean. All fan wheels
and propellers are balanced at the factory. When they go out of balance
because of the dirt build-up on the blades, it can be quite an annoyance.

■ Check to see if a belt is bad. Maybe the belt has chunks that have
come out.

■ Secure all screws, nuts, and bolts.

■ Check motor mounting and belt tension for flapping belts or slipping belts.

■ Check for loose filters in the filter rack. Metal filters in a metal filter rack will give you noise and vibration.

DIRECTION OF DISCHARGE AND ROTATION OF FANS

It is quite common to see a man take a fan apart and remove the motor only to find he is stumped. The fan is either put together wrong or the fan is running backwards. When you have three shifts, the day shift will pass the job to the swing and there is where the fun begins. See Fig. 19-10.

Shown in Fig. 19-11 are the four most common arrangements for fans. Figure A is called arrangement No. 7. Figure B is called arrangement No. 10. Figure C is called arrangement No. 9. Figure D is called arrangement No. 3.

Arrangement No. 3 is where the motor and driver pulley is under the fan driven pulley and to the right or left side—not directly under. Arrangement No. 7 is where you have direct drive of the fan by the motor. Arrangement No. 9 is where the motor and driver pulley is directly above the fan driven pulley. Arrangement No. 10 is where the motor and driver pulley is directly under the driven pulley.

The rotation of a motor is determined as shown in Fig. 19-11. The shaft side of the motor is the front. Face the front of the motor. If the shaft of the motor is turning in the direction that the hands of a clock would turn as we stand in front of the motor, the rotation is clockwise. If the motor shaft is turning in the opposite direction from the clock, we call this rotation counterclockwise. The abbreviation cw stands for clockwise and ccw stands for counterclockwise.

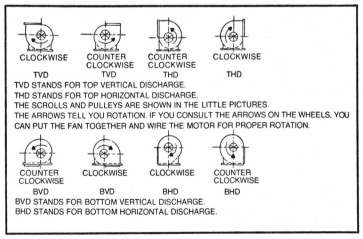

Fig. 19-10. Direction of discharge and rotation of fans.

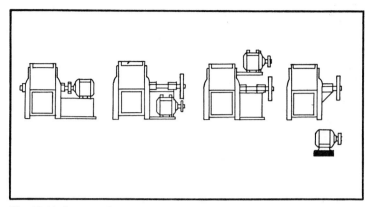

Fig. 19-11. Common fan arrangements.

Other motor positions not shown on the standard motor position drawing could be where motor x driver pulley is directly above the fan driven pulley or motor w driver pulley is directly under fan driven pulley. To do the latter, the fan shaft would be extended. You can have motor x driver pulley directly below the fan pulley and motor w driver directly above the fan pulley. See Fig. 19-12.

The type of fans used in air conditioning, refrigeration and heating are centrifical fans. The class of fans used in air conditioning, refrigeration and heating will be utility or heavy duty industrial. The light duty fan that has thin walled sheet metal scroll is the utility class. The heavy duty fan with plate metal scroll is industrial or commercial class. It is best to consult the fan maker or industrial class fans as there are different grades of industrial fans. The fan maker will use different model numbers for the different grades of industrial fans. In all cases except direct driven fans, the fan driven pulley rpm will be different rpm than the motor driver pulley rpm.

MOTOR DATA FOR FANS

The fan motor tag will show horsepower, power (220, 440, 3600 three-phase or 115- or 230-volts, single-phase), frame size (48, 56, 213,

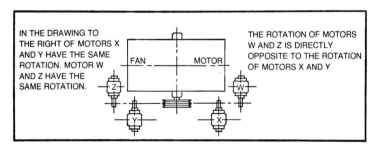

Fig. 19-12. Standard motor positions.

etc.), rpm, service factor (the amount you can multiply times the name plate current to find the maximum current the motor may consume) and type of enclosure. If the end bells of the motor are open on the bottom, this type of motor is called a drip-proof motor. When the motor has no opening and there is an external fan with a little cover over it so that the air is directed over the fins on the motor case, this type of motor is called totally enclosed fan cooled.

DRIVE DATA FOR FANS

There are direct driven fans, fans with a variable pitch and fixed pitch pulley, or fans with fixed pitch pulleys. The driver pulley is on the motor and has a diameter. The driven pulley is on the fan and has a diameter. Most of the time the variable pitch pulley will be on the driver as for the most part of the driver pulley is smaller pulley and the smaller pulley is the less expensive pulley. With the exception of the knotched belts, the rest of the belts for the pulleys will be V belts.

Fan units have vibration isolators. The fan can be supported or suspended by springs, rubber pads, or a combination of spring or pad. Fan units have vibration eliminators. Where the return air, discharge duct, or hot and cold deck are attached on the fan, you will find a canvas coupling 6 to 9 inches wide between the fan and the above ducts or decks. The purpose of the canvas is to keep fan noise from being transmitted by duct or deck into the rooms of the building. Check the canvas every year because the canvas will rot and you will loose cfm through the rotten canvas holes.

FAN SUPPLY AIR

The average building heating and air conditioning fan system will have a supply air that will be approximately 20 percent outside air and 80 percent return air. This relationship can be changed by a variable damper that is located in the outside supply air of the fan unit.

The simplest fan system for heating and air conditioning is a single zone as shown in Fig. 19-13. We combine outside air with return air and the air travels through the filter system, hot and chill water coils, the fan section, the duct network into the rooms. The temperature controls are air. We provide control air to a thermostat in the room and the thermostat branch line feeds both the hot and chill water coils. The air thermostat equates temperature to air pressure and throttles the three-way valves of the hot and chill water coils. The spring in the heating three-way valve is 2 to 7 psi and the spring in the chill water three-way valve is 4 to 11 psi. The thermostat branch pressure/temperature relationship goes like this.

Branch pressure	Temperature
0 psi	Full heat. Hot water coil open, cooling coil closed.

As we go from 0 to 7 psi the heating will diminish.

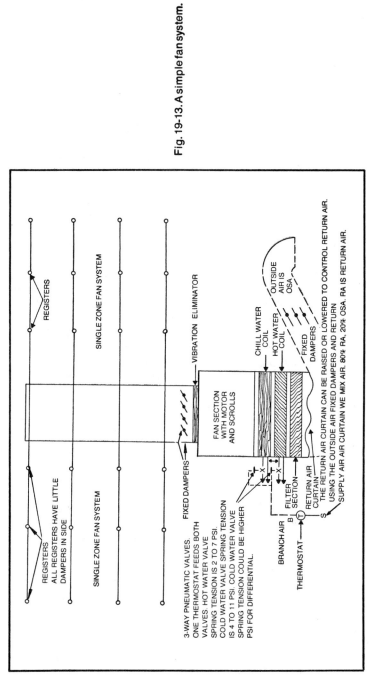

Fig. 19-13. A simple fan system.

REGISTERS

SINGLE ZONE FAN SYSTEM

VIBRATION ELIMINATOR

CHILL WATER COIL

HOT WATER COIL

OUTSIDE AIR IS OSA

FIXED DAMPERS

FAN SECTION WITH MOTOR AND SCROLLS

THE RETURN AIR CURTAIN CAN BE RAISED OR LOWERED TO CONTROL RETURN AIR.
USING THE OUTSIDE AIR FIXED DAMPERS AND RETURN
SUPPLY AIR AIR CURTAIN WE MIX AIR. 80% RA, 20% OSA. RA IS RETURN AIR.

RETURN AIR CURTAIN

FILTER SECTION

B

THERMOSTAT

BRANCH AIR

S

REGISTERS
ALL REGISTERS HAVE LITTLE
DAMPERS IN SIDE

SINGLE ZONE FAN SYSTEM

FIXED DAMPERS

3-WAY PNEUMATIC VALVES.
ONE THERMOSTAT FEEDS BOTH
VALVES. HOT WATER VALVE
SPRING TENSION IS 2 TO 7 PSI.
COLD WATER VALVE SPRING TENSION
IS 4 TO 11 PSI. COLD WATER VALVE
SPRING TENSION COULD BE HIGHER
PSI FOR DIFFERENTIAL.

157

Branch pressure	Temperature
8 psi	We are one-half heating and one-half cooling (approximately).
9 psi	No heating and cooling is starting.

From 10 to 16 psi the cooling increases.

10 psi	The cooling coil is just starting to come on the line. Heating coil is off.
16 psi	The cooling coil is 100 percent on the line.

The outside and return air is tempered by both coils (heating and cooling) which have been throttled by the air thermostat/three-way valves and the air is delivered into the zone area where the thermostat is sensing temperature.

Figure 19-14 shows a multizone fan system with hot and cold decks. The outside air and return air are mixed and travel through the filters, fan section, fan air plenum and through the two hot water coils. There is a hot water coil on the left side of the fan air plenum and a hot water coil on the right side of the fan air plenum. After the air leaves the hot water coils it is hot and fills up the right and left hot decks. The remaining air in the fan air plenum will travel through the chill water coil and fill the cold plenum and entire cold deck. We leave the hot and cold decks by way of a zone that has dampers which will blend the two air streams into a duct which goes to the room area where the thermostat is located. The zone dampers that blend the hot and cold deck air are linked together so that when one damper closes the other will open. We have the same relationship of thermostat branch air pressure to heating and cooling dampers as with the thermostat and hot and chill water coils (substitute dampers for coils).

Figure 19-15 shows the air controls for water and air temperature of a typical multizone fan system with hot and cold deck. Colored tubing runs between the junction points. We have a transmitter in the return air (black tube), transmitters for hot decks (clear and yellow tubes), and a transmitter for the cold deck (lavender tube). There is a three-way valve for the cooling coil, a three-way valve for the heating coil left side, and a three-way valve for the heating coil right side.

A transmitter is an air control that will relate temperature into air pressure. The difference between a transmitter and thermostat is that a transmitter can only relate temperature into pressure, but cannot control air valves or air motors like the air thermostat. A transmitter has no set point. They have a range of 0 to 100 plus degrees F.

When you look at the pneumatic thermometer in Table 19-1, you can read 58 degrees F or 10 psi. The transmitter works in the following manor. Feed a small restrictor device with 20 psi supply air. On the arrow side of the restrictor, feed a transmitter. The transmitter will bleed off a small amount of air until we reach a certain pressure on tubing length A that is proportional to the temperature that the transmitter is sensing. See Fig. 19-16.

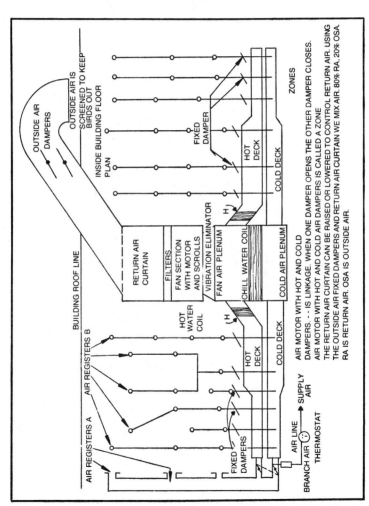

Fig. 19-14. A multizone fan system with hot and cold deck.

OUTSIDE AIR DAMPERS

OUTSIDE AIR IS SCREENED TO KEEP BIRDS OUT

INSIDE BUILDING PLAN

FIXED DAMPER

HOT DECK

COLD DECK

ZONES

BUILDING ROOF LINE

RETURN AIR CURTAIN

FILTERS

FAN SECTION WITH MOTOR AND SCROLLS

VIBRATION ELIMINATOR

FAN AIR PLENUM

CHILL WATER COIL

COLD AIR PLENUM

H

HOT WATER COIL

AIR REGISTERS B

AIR REGISTERS A

H

HOT DECK

COLD DECK

FIXED DAMPERS

AIR LINE

BRANCH AIR

THERMOSTAT

SUPPLY AIR

AIR MOTOR WITH HOT AND COLD DAMPERS. - - IS LINKAGE. WHEN ONE DAMPER OPENS THE OTHER DAMPER CLOSES. AIR MOTOR WITH HOT AND COLD AIR DAMPERS IS CALLED A ZONE THE RETURN AIR CURTAIN CAN BE RAISED OR LOWERED TO CONTROL RETURN AIR. USING THE OUTSIDE AIR FIXED DAMPERS AND RETURN AIR CURTAIN WE MIX AIR. 80% RA, 20% OSA. RA IS RETURN AIR. OSA IS OUTSIDE AIR.

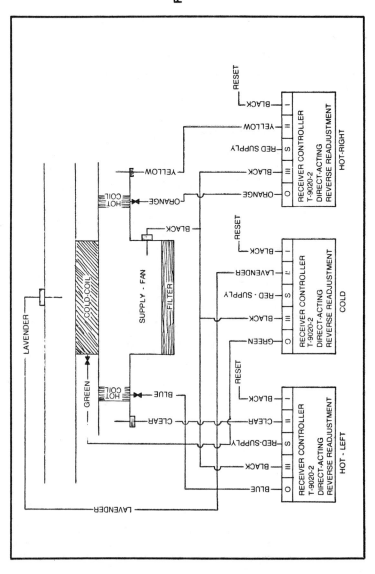

Fig. 19-15. Air controls.

Table 19-1. Pneumatic Thermometer.

Degrees F	Pressure psi
0	3
9	4
16	5
25	6
33	7
42	8
50	9
58	10
67	11
76	12
83	13
92	14
100	15

Table 19-2. Hot and Cold Air Temperatures.

OSA	HC	CC	OSA	HC	CC
40	110		76	86	59½
41	109⅓		77	85⅓	59
42	108⅔		78	84⅔	58½
43	108		79	84	58
44	107⅓		80	83⅓	57½
45	106⅔		81	82⅔	57
46	106		82	82	56½
47	105⅓		83		56
48	104⅔		84		55½
49	104		85		55
50	103⅓		86		54½
51	102⅔		87		54
52	102		88		53½
53	101⅓		89		53
54	100⅔		90		52½
55	100		91		52
56	99⅓		92		51½
57	98⅔		93		51
58	98		94		50½
59	97⅓		95		50
60	96⅔		96		49½
61	96		97		49
62	95⅓		98		48½
63	94⅔		99		48
64	94		100		47½
65	93⅓	65	101		47
66	92⅔	64½	102		46½
67	92	64	103		46
68	91⅔	63½	104		45½
69	90⅔	63	105		45
70	90	62½			
71	89⅓	62			
72	88⅔	61½			
73	88	61			
74	87⅓	60½			
75	86⅔	60			

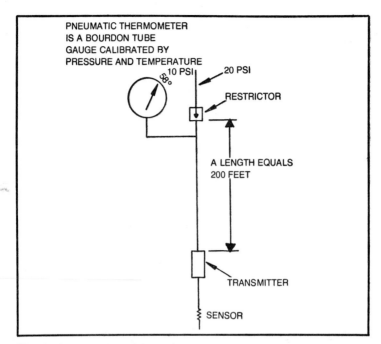

PNEUMATIC THERMOMETER
IS A BOURDON TUBE
GAUGE CALIBRATED BY
PRESSURE AND TEMPERATURE

Fig. 19-16. How a transmitter works.

The transmitter will bleed off more than the restrictor can supply the tubing section of line A. The transmitters will go to number 11 on the receiver controllers. The receiver controllers will throttle the three-way valves for the two hot coils and cold coil. The transmitter provides the air signal to the receiver controller and the controller does the work. Table 19-1 shows the temperature/pressure relationship for all transmitters. The water temperature and air temperature in the hot and cold decks are inversely proportional to the outside air. The receiver controller computes this relationship with the help of the transmitters and pneumatic three-way valves.

Table 19-2 will give you hot and cold deck air temperatures for the outside air temperature. Example: with an outside air of 70 degrees F, the air in the hot deck would be 90 degrees F and the air in the cold deck would be 62½ degrees F. If the outside air is 80 degrees F the air in the hot deck would be 83⅓ degrees F and the air in the cold coil would be 57½ degrees F.

The crossover for heating is 82 degrees F at 82 degrees F outside air (the hot deck air is 82 degrees F). The crossover for cooling is 65 degrees F. When the outside air is 65 degrees F, the cold deck air is 65 degrees F. A pneumatic thermometer can be placed anywhere along the transmitter line for the hot and cold decks and you can read the temperatures and compare the deck temperature to the chart temperature.

Chapter 20
Humidity

Humidity is the amount of water vapor or moisture that is in the air compared to its total saturated state of water vapor per given temperature. When we measure air temperature, the air has two temperatures. There is a dry bulb air temperature and a wet bulb air temperature. The dry bulb air temperature is a measurement of sensible heat that our bodies can feel. This temperature is measured by a thermometer such as the kind used outside a home and shaded from direct sunlight. The wet bulb air temperature is a measurement of the latent heat of vaporization. Latent heat of vaporization is a heat that exists in the air that the human body cannot sense.

The wet bulb thermometer has a wick or sock over its bulb that is wet with water. The wet bulb thermometer will normally read a temperature that is lower than dry bulb thermometer. The temperature you read off the wet bulb thermometer is of the air that passes over the wick that is boiling off (evaporating) the moisture of the water of the wet wick to give moisture vapor temperature.

When dry and wet bulb temperatures of the air are far apart, the humidity is low. As the dry and wet bulb temperatures come together, the humidity will rise. When the dry and wet bulb temperatures are the same, the relative humidity is 100 percent (dew point). The air will start giving up some of its moisture content as it is saturated with moisture at 100 percent relative humidity. A drop of 1 degrees F will force the air to give up moisture. When the dry and wet bulb are very close on temperature, the ground temperature, if it is lower, will cause the air to give up little droplets of moisture because the air cannot hold the moisture beyond saturation point.

Most water towers at 85 percent relative humidity are at their top limit. When the building gets too warm and reaches 85 percent relative humidity, you will have to put more tonnage on the line. This means using all tower pumps, chill water pumps, and compressors. In the case of air-cooled condensers, you can increase efficiency by wetting down the condenser.

RELATIVE HUMIDITY

When the building relative humidity becomes too high (general building air conditioning is 50 percent) because of the heat loads and people loads, you can close off the outside air and place the system on 100 percent return air. The refrigeration coils on the second through .4 passes will dry the air out to 50 percent relative humidity or less, plus lower the temperatures. Also, increase the hot and chill water temperatures and increase the hot and cold deck temperatures. Chill the air maximum to wring out the moisture and reheat maximum. Use a sling psychrometer inside and keep metering the building air until you reach the 50 percent relative humidity. When the building reaches 50 percent relative humidity, start adding more outside air and cut back on return air. You do this by fixed dampers on curtains on outside and return air.

When the humidity is too low, fine furniture and wood materials will crack and could be destroyed (Table 20-1). Set a large pan of water in front of the filter section of the fan system. You can use portable humidifiers, open pans placed around the area, water boiled off in the room as steam or get a swamp cooler and set it in the area of the building that has too low a humidity.

Swamp coolers are directly affected by relative humidity. When the outside relative humidity is over 80 percent, you might as well turn off the swamp cooler because you are not doing anything but moving dust and air that is fully saturated with moisture. If the building is cool, 65 to 67 degrees F, raise the humidity to 80 percent relative humidity and the people will feel warm and comfortable. Use the building humidifiers, steam sprays, water slingers, or wheels and swamp coolers to raise the humidity.

If the building is too hot, 80 degrees F, drop the humidity below 50 percent, if possible. Use the dampers and curtains for outside and return air. Increase hot and cold deck temperatures, hot and chill water temperatures, and chill air maximum to wring it out of moisture and reheat maximum.

READINGS

It is a good idea to take a relative humidity reading of the building and outside air one or more times a day. When working in the storage field

Table 20-1. Humidities for Storage.

BOOK STORAGE	50%
FLOUR STORAGE	65 + OR −10%
FUR STORAGE	55%
GENERAL BUILDING AIR CONDITIONING	50%
GENERAL STORAGE	50 + OR −10%
GRAIN STORAGE	42 + OR −20%
GREEN HOUSES FOR TROPICAL PLANTS	80%
LABORATORY CLEAN ROOMS	33 + OR −10%
MEAT STORAGE	87 + OR −5%
PRODUCE	87 + OR −5%
PHARMACEUTICALS	50% OR LESS
PHOTOGRAPHY	50%
RUBBER	50%
TEXTILES	65%
TOBACCO	75%

where humidity is critical, take the readings frequently. There are many types of hygrometers on the market. Some have charts that a stylis will scribe the dry and wet bulb or give the reading directly in percent relative humidity. Others have a face plate and a needle that will give you the relative humidity.

■ There is a time delay or lag time associated with all of these expensive meters. The reading that you see on the chart of face plate might be over an hour old. It could be more time has lapsed. The response time is giving you old information. Manufacturers will not tell you that their meter has response time or that you will have to wait a long period of time to see what the relative humidity was.

■ You will have to send all these expensive meters out for calibration, maintenance, and repairs at least every two months.

■ When you are dealing with a process or storage situation where the humidity is critical and a lot of money could be lost because of the wrong humidity, you want a meter that will give very high accuracy and give it in three minutes or less. The Bacharach Sling Psychrometer is the only way to go. If you want a permanent record, then go out and spend a chunk of money and purchase a chart-recording hygrometer.

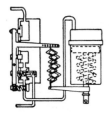

Chapter 21
Full Hermetic
Compressors

A full hermetic compressor is a compressor of which the motor is a permanently sealed integral unit. The word *Hermetic* means sealed in unit. To open a sealed compressor, use a portable grinding wheel and grind away the weld bead, at point B of Fig. 21-1, all around the can. It will take over 20 minutes. Do not use a cutting torch as everything inside of the can will be set on fire because of the high heat. There is about a pint to a quart of oil in the bottom of the can.

Letter "C" of Fig. 21-1 is the suction or low side of the compressor. The large pipe is suction. There are some hermetic compressors that have the can laying on the elongated side and the welding bead of the two halves will be horizontal.

PROCESS TUBE

Letter D of Fig. 21-1 shows the process tube. The process tube is a little pipe the compressor manufacturer will put on the can so that the refrigeration and air conditioning manufacturers and men that work the trade can gain access on the low side so that the unit can be charged or vacuumed without the cost of a service entrance valve. The procedure I use is to install a male one-fourth flare fitting on a copper tube about 4 inches long. Then silver solder the flare tube on the process tube. Next, pull your vacuum and final charge the unit.

After final charge, a pinch-off is placed on the process tube. A *pinch-off* is a small wedge vice with wing nuts on either end that you tighten to pinch the process tube and valve off the service entrance. A vice grip that has wide jaws could be used. Keep the pinch-off on the process tube and take a side cutter or hacksaw and cut off the flare tube.

Next, flatten and bend over the process tube where you have made the cut. Take a torch and silver solder the flattened end to give a permanent seal and remove the pinch-off. You must not remove the pinch-off until after the flattened process tube has been silver soldered. A tiny little bit of gas escaping through the flattened process tube will cause a leak and a job that will have to be reworked within the week.

The refrigeration trade has a special tool kit called a *process tube tool kit*. It is a one-fourth inch male flare with rubber bushing and nut that can be attached to the process tube and removed by tightening or untightening the nut and rubber bushing. Because I have to use a torch to seal the end of the process tube, I just make my own process tube tool kit and save about $30.00. Letter E of Fig. 21-1 shows the discharge pipe. It goes to the condenser, reversing valve, or hot gas bypass.

Letter E, Fig. 21-1 is high side and the discharge is normally smaller than the suction pipe. It is located in the bottom half of the can on 95 percent of the hermetic units. On small hermetics such as one-eighth or one-fourth horsepower, the discharge and suction pipes are the same size. When in doubt, use my rule of the pipes: in the top half of the can is the suction and the pipe and in the bottom half is the discharge. To be sure, you can start up the compressor as you see it in Fig. 21-1 and put your hand over the pipes.

If there is more than one pipe in the bottom half of the can, the two that are not discharge are oil cooler pipes. They are attached to a small tube and fin coil or a small static coil.

TERMINAL BOX

Letter F of Fig. 21-1 shows a cover for the motor terminal box. The three letters that are on the outside of this cover stand for common, start, and run.

Letter G of Fig. 21-1 shows a pin and bake-e-lite motor terminal with little spades on the ends of the pins. The pins on the inside of the can have little spades on them and the wires are attached to the spades by little stay-com lugs. Motor terminals can also be nut and bolt. This motor terminal could be either single- or three-phase, 120/480 volts. There is a name plate on top of the can that gives this information and more.

Letter H of Fig. 21-1 shows the feet of the compressor. There can be rubber grommets that attach to these feet or the feet can rest on springs. Most compressors have either three or four feet and many times when you replace a compressor you must buy a special adapter plate to transition the feet of the new compressor which are spaced different to the mounting hardware of the old compressor that is in the condenser unit.

In Fig. 21-2, the top half of the can has been removed and placed to the right side. The wires that go to the stator winding of the motor were pulled off the spades of the pin and bake-a-lite motor terminal on the inside of the can and are hanging outside of the bottom half of the can. As you look inside the bottom half of the can, you will see a pint or more of refrigeration oil

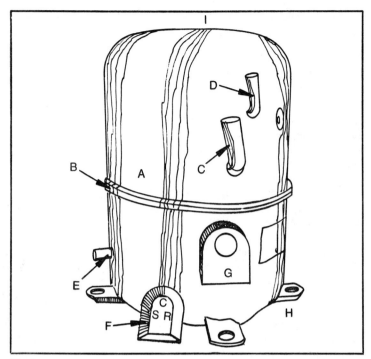

Fig. 21-1. A sealed compressor.

and you are looking at the inside of the crankcase of the compressor. On the outside of the bottom one-half of the can, you will find a crankcase heater strapped around the can. You will find the heater strapped on the outside bottom of the crankcase or inserted in a little well that extends into the bottom of the crankcase. The purpose of the heater is to prevent the liquid refrigerant from settling in the crankcase of the compressor.

VAPOR PUMPS

The reed valves of the compressor can be broken by liquid refrigerant. All refrigeration compressors are vapor pumps. The valves (with the exception of centrifical, screw, and rotary compressors which have no valves) are fluttering reed valves. The valves are operated by vapor pressure. Any large amount of liquid refrigerant can break them.

On 90 percent of all systems, the compressor is the lowest spot on the system and liquid will run and settle in the lowest spot. However, liquid refrigerant does not like heat and will not settle in the crankcase if there is a heater there. The wattage of the heaters are 20 to 60 watts.

Letter A in Fig. 21-2 shows the suction pipe which enters the can and dead-ends. The entire open space inside the can is low side or suction pressure. The windings of the stator are cooled by the suction return refrigerant vapor.

Letter B of Fig. 21-2 shows the process tube.

Letter C of Fig. 21-2 shows a thermo overload switch (bimetalic) which will open when the winding in that area of the stator winding becomes too warm and will close when the winding of the stator cools down. The wire that attaches to the thermo overload switch is the common wire.

Letter D of Fig. 21-2 shows the stator winding. If you have two windings in the stator, you have single-phase. Three windings in the stator is a three-phase motor.

The motor in Fig. 21-2 was a burn-out. A hot spot developed and the wires cooked and shorted to each other before the switch could sense and open the circuit. The windings of the stator are cooled by the suction return refrigerant vapor. The only time the stator winding is not refrigerant-cooled is when you have a stator that is pressed into the top half of the can or there are fins that are mounted on the top half of the can. In this case, the compressor manufacturer is relying on outside air to cool the stator winding.

Letter E of Fig. 21-2 shows a suction muffler that is piped to the suction of the compressor. Letter F of Fig. 21-2 shows the motor terminal cover with the letters C, S and R that indicate the location of common, start, and run. Letter G of Fig. 21-2 shows the motor terminal.

Figure 21-3 is another view of the hermetic compressor. Letter A of Fig. 21-3 shows the suction muffler with two pipes going to the two suction valves. Letter B of Fig. 21-3 shows the discharge muffler. We have one muffler that is common to two discharge valves. Letter C of Fig. 21-3 shows the discharge pipe that is curved and long so that it will not become fatigued and broken by vibration. Letter "D" is the discharge pipe extended through the bottom half of the can.

Figure 21-4 shows another view of the compressor (without the stator winding). Letter A of Fig. 21-4 shows the suction muffler. Letter B of Fig. 21-4 shows the discharge muffler. Letter C of Fig. 21-4 shows a discharge pipe. Letter D of Fig. 21-4 show a discharge pipe. Letter E of Fig. 21-4 shows the rotor of the motor. Letter F of Fig. 21-4 shows fan blades mounted on rotor to help cool windings of stator with refrigerant vapor.

ROTOR AND STATOR

Figure 21-5 is a rotor and Fig. 21-6 is a stator. Letter C in Figs. 21-5 and 21-6 shows the little fan that will whisk the vapor refrigerant around the windings of the stator. The motor compressor unit is suspended by springs. There is very little chance of breaking loose a locked rotor by using a block and hammer on the outside can of the compressor. The stator winding has little stay-com spades (D of Fig. 21-5) that clip to the pin and bake-a-lite motor terminal inside the bottom can.

We do not have a starting switch for the start winding when a compressor like this is single phase. The start winding of a full-hermetic

Fig. 21-2. A hermetically sealed compressor.

compressor is switched on the outside of the can by a current, potential, hot wire, or solid-state relay. The only exception is a permanent-split, capacitor-type motor. The run capacitor is permanently connection between start and run on the outside of the can.

Figure 21-7 shows the top half on the can (A) and letter B is the bottom half of the can. Letter C is the springs that support the compressor and motor unit inside of the can. Letter D is motor terminal spades inside the can. Letter E is the inside of the crankcase.

Figure 21-8 shows the discharge muffler (A), the valve plate (B) the discharge reed (C), and the valve reed stop (D). The reed stop will keep

Fig. 21-3. A compressor.

the discharge reed from fluttering too far and breaking off. These reeds have the finest of steel; otherwise they would fatigue and break off. Reed valves are operated by gas pressure only.

In Fig. 21-9 letter A is the suction muffler. Letter B is the discharge muffler. Letter C is the discharge pipe. Letter D is the suction pipe.

Figure 21-10 shows the compressor without the stator and mufflers and shows the valve plate with suction reed valves. Letter A is the rotor. Letter B is the fluttering suction reed valves. Letter C is the suction valve ports (two ports per reed valve). Letter D is the cylinder of the compressor. Letter E is the compressor mounting studs that attach to the springs inside of bottom half on the inside of the crankcase. Letter F is the studs to mount the stator.

Figure 21-11 shows valve plate with suction reed valve (letter A).

Figure 21-12 shows the other side of the valve plate. Letter A is fluttering reed discharge valve. Letter B is discharge reed valve stop. Letter C is the valve plate.

In Fig. 21-13, letter A is the cylinder. Letter B is piston (piston does not have rings, but has grooves. Letter C is the rod and rod bearing. Letter D is the rotor.

In Fig. 21-14 letter A is the crankshaft. Letter B is the piston. Letter C is the rod shim washers. Letter D is the rod. Letter E is the half-moon brass rod to crankshaft bearing. Letter F is the piston to rod bearing.

Fig. 21-4. A compressor.

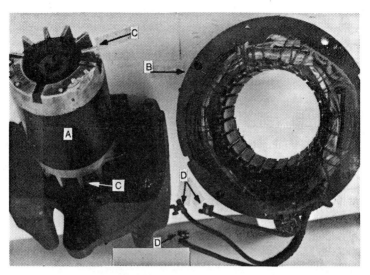

Fig. 21-5. A compressor rotor.

COMPRESSOR REPAIR

When I was a teacher, I used to take the class on a field trip to a compressor rebuilding plant. They used a round rotating table on top of which the compressor was clamped on by its legs and C clamps. There was a grinding wheel that was pushed against the weld bead on the can of the compressor by spring tension. It took about 10 minutes to grind away the weld bead and open the compressor. Care must be taken to remove the weld bead, but you must keep as much as possible of the flanges of the two cans so that they can be rewelded.

The owner of the shop would place all the opened compressors in groups of the same type on the large floor area of his shop building. When a customer purchased a rebuilt, he had to give an exchange or broken-down compressor. Compressors came in with broken valves, discharge lines, frozen bearings, broken pistons and rods, and burned out stators. With 80 percent of the compressors rebuilt, the shop owner could use the parts from the compressors sitting in groups on the floor. Once in a while, a stator was sent out for rewind.

Every once in a while, a compressor would come in and nothing was wrong with it. The problem was a possible defective starting relay, capacitor, control problem, incorrect voltage, or bad klixon.

After the defective parts are replaced, the compressor, motor, and cans are cleaned with a non-flammable solvent—or soap and water—and dried out. The compressor is placed in the cans, the condensor pipe is joined in the can, the motor electrically hooked up inside the can, and everything is assembled.

173

The top and bottom of the compressor cans are clamped together by three or four C clamps. Using electric arc welding, the cans were tacked in about seven or eight places equal distant on the flange of the cans. After tacking, the C clamps were removed and a bead was welded on the cans by electric arc. The tac weld in eight places was to keep the flanges lined up and prevent warping because of the heat.

Electric arc is the only welding permitted because we can get a good penetration with minimum heat on compressor parts and electric motor stator. These compressor parts must have no flammable residue on them.

Fig. 21-6. A compressor stator.

Fig. 21-7. The top and bottom of the compressor can.

Rubber corks are placed in the suction and discharge pipes. A nitrogen or Freon hose is clamped on the compressor process tube and the unit is filled with gas (100 psi pressure). While the hose is still attached, the compressor was submerged in a large water tank and the shop foreman would check for bubbles and a possible leak in the can weld. If a leak occurred, it was marked and arc welded again.

If we passed the leak test, the hose was removed from the compressor process tube. The proper amount of refrigeration oil was

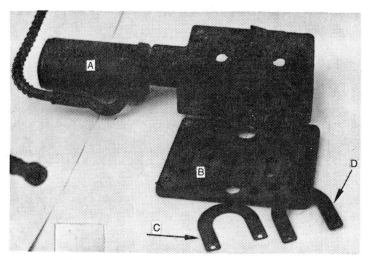

Fig. 21-8. Compressor parts.

poured into the compressor using a funnel in the suction pipe. The suction was corked. The nitrogen hose was placed on the process tube and the compressor can filled with nitrogen. The hose was removed and a cork placed on the process tube before too much nitrogen gas could escape. Nitrogen gas is used to pressurize the compressor can because it will help dry up any moisture inside the compressor can. The next step is that the compressor is painted black. A bill and guarantee is attached and the rebuilt compressor is placed in inventory for a customer.

The largest size on most hermetics is 7½ horsepower, three-phase. However, General Electric makes a line of full hermetics that exceeds 10

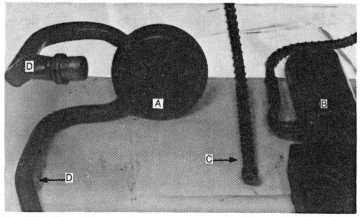

Fig. 21-9. Compressor parts.

Fig. 21-10. Compressor parts.

horsepower, three-phase. Many air conditioning systems will use three hermetic compressors in the compressor condenser section.

The proper charge of refrigerant in a refrigeration system is critical to a full or semihermetic compressor. The cooling of the stator winding is done by the refrigerant. When you are short of refrigerant in the system, you are asking for a burnout stator.

Too much refrigerant will cause the system to have too high head pressure and overwork the motor—then along comes another burnout. It does not take long to develop a hot spot or to cook the stator winding. A dirty or plugged filter will cause too high head pressure, too low suction, and a good chance of burnout stator. There is a chance of liquid flooding

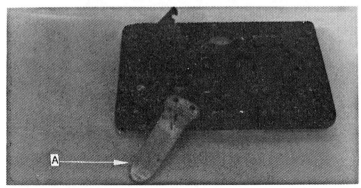

Fig. 21-11. Valve plate.

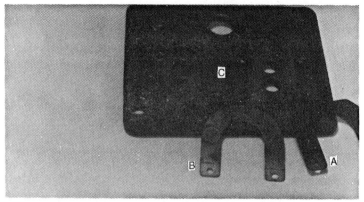

Fig. 21-12. A valve plate.

into the suction of compressor because of low pressure, low temperature. Liquid flood-back will break the fluttering reed valves.

COMPRESSOR REPLACEMENT

Reasons to replace compressor are:

- The pressures on suction and discharge are the same—bad valves.
- The ammeter reads nameplate locks rotor currents—locked rotor.

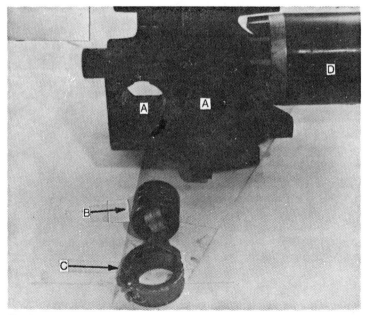

Fig. 21-13. A cylinder.

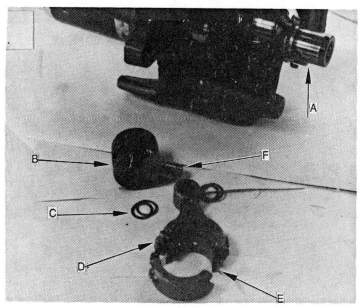

Fig. 21-14. Compressor parts.

■ Short of motor common, start and run terminal to shell or compressor can—burnout.

■ Open winding—burnout.

■ Incorrect ohmage of motor windings. Same ohmage across all motor terminals—burnout.

If you have not observed problems in the proceeding steps the trouble is not the compressor. Look elsewhere:

■ Electrical malfunction.

■ Obstruction in the refrigeration system.

■ Faulty expansion device.

■ Incorrect refrigerant charge.

■ Dirty filters.

Figures 21-15A and 21-15B are different angles of the same automotive compressor. The compressor is semihermetic because of the bolts and it has an outside drive. A semihermetic compressor is one that is of nut and bolt construction and can be taken apart for repairs. There might be an electric motor inside the case or the semihermetic compressor might have an outside drive. An outside drive is where the compressor shaft extends through the case by way of a seal and is driven by an electric motor or engine either by belts and pulleys or direct coupling.

Letter A of Fig. 21-15A is a Schrader valve port (service entrance) that is on the high side. This is the high side because the hose (C) that port

Fig. 21-15A. An automotive compressor.

A is on goes to the condenser that is in front of the radiator. The condenser is on the high side. Service entrance valve port B is on the low side as it is going toward the inside cabin of the car. Automotive and truck refrigeration is the only place where you will find rubber hose piping.

The piping that attaches to compressors can either be welded or held on by threaded nuts with little plastic "O" rings (see point E). These nuts

Fig. 21-15B. An automotive compressor.

are called roto-locks. Letter F is the clutch of this compressor. This is the movable part of the clutch assembly.

Letter G is the other half of the clutch with the pulley. Between F and G on the inside of this assembly there is a magnetic clutch coil that is mounted to the compressor case. One side of the coil attaches to wire H which goes to the thermostat inside of the car and the other side of the magnetic clutch coil goes to the compressor shell which is the car ground. When there is no cooling inside of the car and the clutch (F) is not turning with G, tape a knife and strip wire H back a little and place a jumper between H and plus battery. If the clutch engages, the problem is with the overload fuse or controls (thermostat or outside or inside sensors).

Letter D is the shroud or fan concentrator. The shroud helps the radiator fan to develop draft so that more air can be moved through the condenser and radiator.

There is one problem with the service valves on this system. If you have to replace the compressor, the system cannot be valved off and the compressor replaced and the system valved on. On this affair, you have to blow charge, replace the compressor, vacuum and charge. When you have the three-way valves, you valve off the system, replace the compressor, valve on system and you are in business.

When a hose breaks cut the hose clean on either side of the break. Insert a piece of copper tubing into each side of the cut hose about 2 inches as a coupling. Take two hose clamps and tighten the cut hose ends onto the copper tubing, vacuum and charge system.

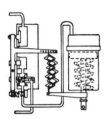

Chapter 22

Expansion Valves

An expansion valve is a device used in refrigeration to drop a high-pressure refrigerant liquid to a low-pressure refrigerant vapor. The expansion valve is fed by high-pressure refrigerant liquid from the condenser. The valve drops the high pressure of the liquid refrigerant and feeds the evaporator with a low-pressure refrigerant vapor. Expansion valves will only drop pressure of liquid refrigerant. If fed with vapor, the valves will not function properly. There are two types of expansion valves: thermostatic and automatic.

THERMOSTATIC EXPANSION VALVES

Super heat is a term used with expansion valves. *Super heat* is the difference in temperature between the refrigerant vapor that flows through the evaporator and temperature of the skin or outside metal of the evaporator. Air conditioning in general has a super heat of 10 degrees. This means that the refrigerant temperature in the evaporator, which is 40 degrees, will give a surface temperature of the evaporator of 50 degrees (or a difference of the 10 degrees which is 10 degrees super heat). Refrigeration used in different fields or processes might have a slightly different super heat.

All expansion valves have a spring inside them that will maintain a super heat (known as the super heat spring). The super heat spring tension can be adjusted by a screw to raise or lower the super heat of the valve. The thermostatic expansion valve is controlled by the power head. The power head is composed of an enclosed diaphragm, a capillary tube, and a sensor bulb with gas charge.

When the sensor bulb becomes warm, the gas charge becomes warm, the gas charge inside the bulb will expand, run inside the capillary tube, and push out on the diaphragm. If the sensor bulb is cold, the gas charge inside the bulb will contract and some of the gas will leave the diaphragm by way of the capillary tube and return to the bulb in this contraction process. When this happens, the diaphragm will contract inside the power head. The power head sensor bulb is normally clamped within 8 inches of the end of the evaporator coil on the suction line to sense the super heat of the coil.

When the sensor bulb that is clamped on the suction return—and as close as possible to the evaporator—becomes warm, the gas charge inside of the bulb expands, runs inside the capillary tube and forces the diaphragm inside the power head out in the direction of force P1. See Fig. 22-1.

When the valve pins move in the direction of P1, the valve needle is moved away from the valve seat. Liquid refrigerant enters the seat and sprays around the needle and enters the evaporator as a low pressure particle size spray. This action only happens when the diaphragm force P1 is greater than the super heat spring force P2. A great deal of the time, the needle will find a spot and stay there giving a steady particle size spray because both forces P1 and P2 have equalized each other.

When the sensor bulb becomes too cold, the gas charge in the bulb contracts, the diaphragm contracts, and force P1 becomes weaker than force P2. The valve needle moves closer to the valve seat and less refrigerant moves through the valve.

If the super heat screw is turned in so that the spring exerts more force on P2, the super heat will rise. The gas pressure inside the evaporator will drop and the gas temperature will lower. But when there is less volume of gas in the evaporator, the skin or metal temperature of the evaporator will rise. Super heat works in two directions. The gas temperature lowers because less volume enters the evaporator. Evaporator metal or skin temperature rises because there is less refrigerant to remove heat. Generally, one turn of the super heat screw will change the super heat of the valve 3 to 5 percent.

When you unscrew the super heat screw, force P1 becomes greater than force P2. The power head diaphragm pushes more in on the valve pins and the valve needle is pushed further away from the valve seat. This lets more particle size refrigerant vapor pass through the valve.

Unscrewing the super heat screw lowers the super heat in two different ways. As more refrigerant enters the evaporator and more heat is removed, the evaporator skin or metal temperature becomes colder to touch. The more refrigerant means the gas has higher pressure and higher temperature inside the evaporator.

Do not forget that super heat is the difference between the temperature of the gas in the evaporator and the actual skin or metal temperature of the evaporator. The skin or metal temperature of the evaporator is the temperature you can feel with your hand touching the evaporator. When you want the evaporator to run colder, unscrew the super heat screw one turn and the temperature should lower 3 to 5 percent of the range of the valve. Super heat will become less.

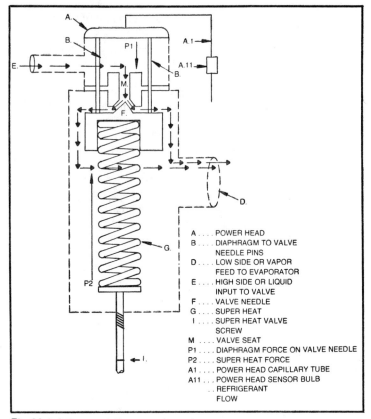

Fig. 22-1. A thermostatic expansion valve.

A .... POWER HEAD
B .... DIAPHRAGM TO VALVE
        NEEDLE PINS
D .... LOW SIDE OR VAPOR
        FEED TO EVAPORATOR
E .... HIGH SIDE OR LIQUID
        INPUT TO VALVE
F .... VALVE NEEDLE
G .... SUPER HEAT
I .... SUPER HEAT VALVE
        SCREW
M .... VALVE SEAT
P1 .... DIAPHRAGM FORCE ON VALVE NEEDLE
P2 .... SUPER HEAT FORCE
A1 .... POWER HEAD CAPILLARY TUBE
A11 ... POWER HEAD SENSOR BULB
        .. REFRIGERANT
        FLOW

If you want the evaporator to run warmer, screw in the super heat screw one turn and the temperature of the evaporator should rise 3 to 5 percent of the range of the thermostatic expansion valve. The super heat will increase.

Be careful when adjusting the super heat screw to lower or raise evaporator temperature. Make note of exactly how many turns and in what direction you have gone. Take it easy. Turn the screw one turn and wait. Have your suction and discharge gauges on the machine when you do this. Suppose that you want a colder evaporator. Unscrew the super heat screw one turn and wait 20 minutes. Repeat the process again. Have a thermometer on the skin or metal of the evaporator. If evaporator pressure becomes too high, condenser pressure too low and bubbles appear in the sight glass, you have too much refrigerant flowing through the valve.

Take it in a turn and wait 20 minutes. Let the system settle down. The valve might search and the pressure not stabilize. You have to reach the proper super heat so that the evaporator will be the coldness desired with no bubbles in the sight glass, no searching of pressure by the thermostatic expansion valve, and condenser pressure high enough to keep refrigerant moving.

Keep notes on how you turn the super heat screw. You have to know where you have been in order to know where you are going. If the pressures have stabilized, but you still have bubbles in the sight glass, you can add a small amount of refrigerant to remove the sight glass bubbles. Observe condenser pressure. Do not exceed the condenser pressures as given in Chapter 9.

Suppose on the other hand, that the evaporator coil is too cold and there is icing on the fins. Turn the super heat screw in a turn and wait 20 minutes. Repeat the process every 20 minutes. It takes 20 minutes for the thermostatic expansion valve to settle down. It is most important to keep your eyes on the sight glass. In addition, observe suction and condensing pressures. The sight glass should give no bubbles unless you turn the valve out instead of in.

When you start starving the evaporator coil, the condensing pressure will become too high and suction drops too low. Be careful. The compressor is refrigerant cooled. Do not burn out compressor. Do not exceed the high side pressures (Chapter 9). The thermostatic expansion valve will automatically adjust refrigerant feed when you have a changing heat load. If the heat load of a system is constant, then the capillary tube, automatic expansion valve, or globe union valve would be more effective and less complicated.

Figure 22-2 shows a typical thermostatic expansion valve. I cut the power head (A) in half and removed the cap tube and sensor bulb. The super heat flange (I), which has the super heat screw, was removed. H is a back for the super heat spring. G is the super heat spring. F is the needle of the thermostatic expansion valve. B is diaphragm to valve needle pins. J is the valve body. E is the liquid high side of the valve. D is the vapor low side. J is the valve body. E is the high side or liquid input for the valve. D is the low side or vapor feed to the evaporator. C is the external equalizer.

All thermostatic expansion valves have either external or internal equalizers. The purpose of an equalizer is to cause the high and low side to balance after the compressor cycles off. The equalizer is a compressor unloading device. When you start the compressor, you will not have the high and low side until you are up to full rpm. Otherwise, the high side could lock up the rotor of the compressor or cause very high starting currents. The internal equalizer will bleed the high pressure to the low pressure until they are equal after the compressor cycles off. This is done automatically inside the valve. The external equalizer has a piece of one-fourth inch tubing that will leave the thermostatic expansion valve at point C and go to the suction return line from the evaporator. When the compressor cycles off, the high pressure is automatically bled off to the low side by this one-fourth inch copper tubing in order to unload the compressor.

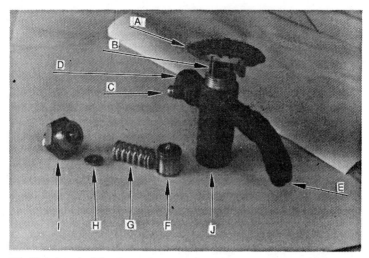

Fig. 22-2. A typical thermostatic expansion valve.

How do we know the high and low side of a thermostatic expansion valve when there is no direction arrow to show refrigerant flow? The small pipe of the valve will be the high side or liquid feed. The large pipe of the valve will be the suction or low side of the valve. The refrigerant flow will be small pipe of valve to large pipe and then to evaporator. When you have a male one-fourth inch flare fitting on the side of the valve body or a small capillary tubing coming off the side of the valve body, this will be the equalizer tube or line going to the suction close to the evaporator. I bring this to your attention because there is one manufacturer that does not put direction arrows on their thermostatic expansion valves.

Figure 22-3 is the same as Fig. 22-2 except that I have turned the super heat screw flange fitting around so that you can see the square screw head. All the other letters are the same.

In Fig. 22-4 I placed the same valve upright and pulled the diaphragm to valve needle pins (B) out so you could see them. The super head spring and its components have been assembled. K is a cover cap for the super heat screw to stop any "would be" leaks.

In Fig. 22-5, I have placed all the movable parts in order for a typical thermostatic expansion valve. A111 is a piece of the diaphragm that is disc-like and it pushes the pins (B) against the needle (F). G is the super heat spring that will push the valve needle (F) and pins (B) against the diaphragm (A111) of power head (A). The two forces that oppose each other are the gas charge in sensor bulb (A11) verses the force of super heat spring (G) which is fixed by super heat screw (I).

Figure 22-6 is a similar thermostatic expansion valve. This valve has the power head and valve body as one cast unit. All other components are the same.

186

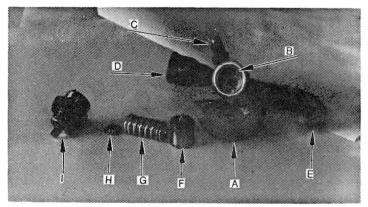

Fig. 22-3. A typical thermostatic expansion valve.

In Fig. 22-7, I have turned the valve so that you can look at the inside of it from the bottom. Note the diaphragm to needle pins (B) and the valve seat (M).

Figure 22-8 is another type of thermostatic expansion valve. The power head (A) is half of the valve body and the super heat screw (1) is on the side. In Fig. 22-9 I have taken the valve apart. Note the valve needle and seat. Figure 22-10 is the same valve without the valve needle, spring, and pin assembly.

In Fig. 22-11, you can see super heat screw and springs from different valves. Letter D is the power head and valve, needle, spring and pin assembly. The valve to the right had two super heat springs.

Figure 22-12 shows a thermostatic expansion valve. It is different from the others because the valve seat is a steel ball (letter A). It is

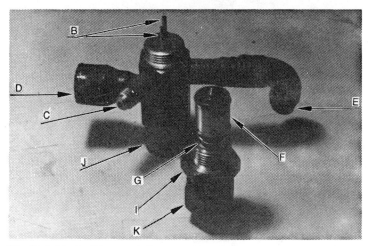

Fig. 22-4. A typical thermostatic expansion valve.

187

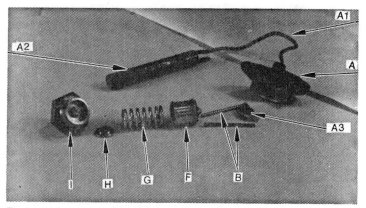

Fig. 22-5. A typical thermostatic expansion valve.

internally equalized. Note that the diaphragm of the power head pushes the valve needle directly with diaphragm to needle pins. The valve seat, ball (A) is spring loaded with the super heat spring. Letter E is the liquid feed high side and D is the low side of this valve.

Figure 22-13 shows a typical thermostatic expansion valve that would be used in cars or trucks for air conditioning. The maximum tonnage is 1½ tons. Letter A is the power head. Letter C is the equalizer tube. Letter D is the low side of the valve. Letter E is the liquid feed or high side of the valve. Letter X is an inline sight glass with the view crystal on top. Letter W is a flexible liquid line hose. Letter Y is a suction line throttling valve.

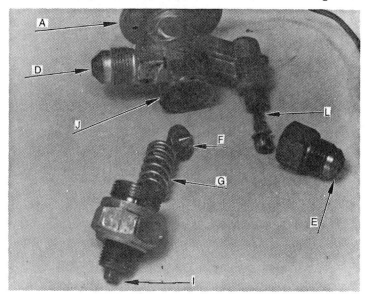

Fig. 22-6. A typical thermostatic expansion valve.

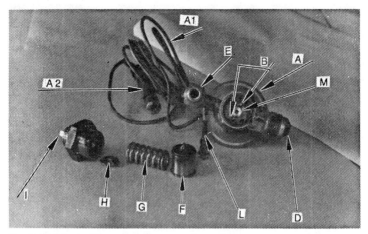

Fig. 22-7. A typical thermostatic expansion valve.

Its purpose is to stabilize the suction gas pressure and not let the pressure temperature of the refrigerant gas get too low. Letter Z is the suction line going to the accumulator and then to the compressor.

An *accumulator* is a device that will trap liquid refrigerant and permit only vapor to pass through and continue on to suction or the low side on the compressor. Accumulators are common on heat pumps and the auto air conditioning field is starting to use them. By trapping liquid, you can extend the life of the compressor valves. Accumulators would be good for low temperature application because they would eliminate flood-back of

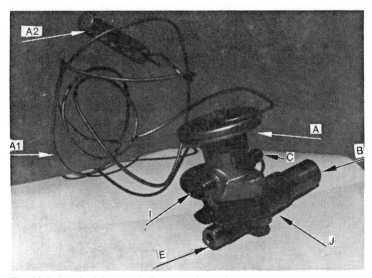

Fig. 22-8. A typical thermostatic expansion valve.

189

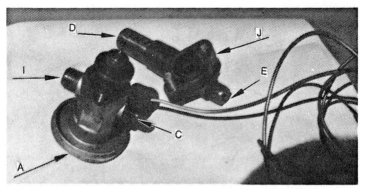

Fig. 22-9. A thermostatic expansion valve.

refrigerant and save the valves. The accumulator lies between evaporator and compressor suction side.

AUTOMATIC EXPANSION VALVES

The automatic expansion valve will change a high-pressure liquid gas to a low-pressure vapor gas. The valves' super heat can be changed manually by either a knob on a super heat screw or the super heat screw alone. It will not adjust to a changing heat load. If the evaporator requires more or less refrigerant, the valve must be manually adjusted.

This valve will work very good on a fixed heat load (Fig. 22-14) is a picture of an automatic expansion valve. Note that the automatic expansion valve does not have a power head with capillary tube and sensor bulb. The

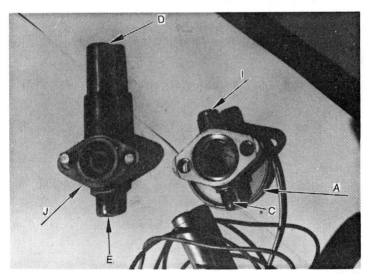

Fig. 22-10. A thermostatic expansion valve.

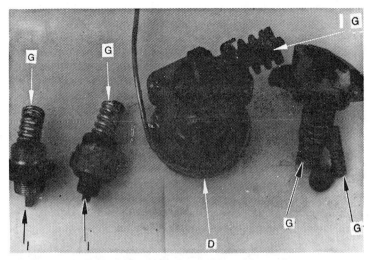

Fig. 22-11. A super heat screw and springs from different valves.

liquid refrigerant enters the automatic expansion valve at pipe E. The high pressure of the refrigerant pushes the needle (F) away from the valve seat (M). The gas volume is throttled by the super heat spring (G) pushing against the needle (F). The low pressure vapor gas leaves pipe D headed for the evaporator. The arrows on in Fig. 22-14 show you the direction of gas flow. Generally, pipe E and D are the same diameter. The high-pressure liquid will normally go to the top half of the valve and the low-pressure vapor will exit from the bottom.

Figure 22-15 shows an automatic expansion valve that has been cut in half and then I cut the top off. Letter I is the super heat screw. Letter G are

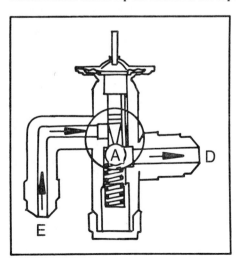

Fig. 22-12. A thermostatic expansion valve.

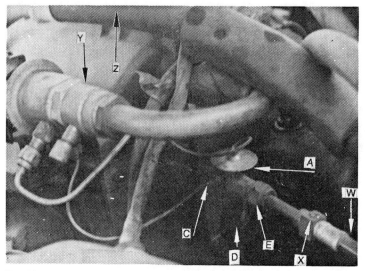

Fig. 22-13. An automatic thermostatic expansion valve.

the super heat springs. This valve had a spring inside of a spring. Letter F is the valve needle (although it does not look like one). Letter J is the valve body. This valve was manually adjusted by a screwdriver in the slot of a threaded disc (I) that did the work of the super heat screw.

One other expansion valve that was used on early types of refrigeration is the Globe Union Valve (or Globe Valve). See Fig. 22-16. This valve has a more convex valve seat (M) and the needle (F) has a triangle. The valve pipes (E and D) are the same diameter and you fed the top half of the valve by pipe E with high-pressure liquid and leave the bottom pipe D with low-pressure vapor. This valve had no springs, moving pins or power

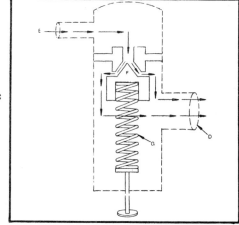

Fig. 22-14. An automatic expansion valve.

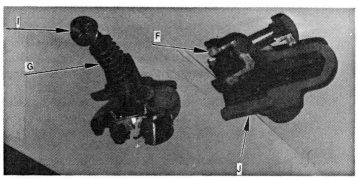

Fig. 22-15. An automatic expansion valve.

heads. All this valve has is a knob (X) that turns stem (Y) to adjust the needle (F) to the valve seat (M). Letter Z is the valve packing. The little arrows show refrigerant flow.

This valve can only be used for fixed-heat loads. However, you can adjust refrigerant flow and by monitoring the evaporator area eventually get the right setting on handle X. When you have a change of heat load, you adjust handle X. If the evaporator area becomes too warm, you open the valve a turn or two turns to get more low pressure cooling refrigerant to the evaporator. When the evaporator becomes too cold, you close the valve a turn or two to have a little less refrigerant flow to the evaporator.

Very little can go wrong with the Globe Valve. Only thing is that this expansion valve requires a fixed heat load and someone to periodically monitor the evaporator area. This valve can be placed on a system and the

Fig. 22-16. A globe valve.

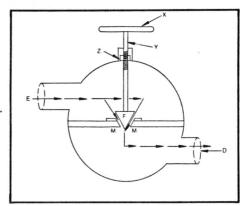

temperature of the evaporator area can be controlled by a thermostat (temperature sensing) and magnetic starter. The Globe Valve has been largely replaced by the cap tube because of price and adjusting problems. People do not want to take the time to adjust the valve to properly feed the evaporator.

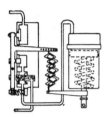

Chapter 23
Capillary Tubes

A capillary tube (cap tube) is a gas pressure dropping expansion device that lies between the condenser and the evaporator. The cap tube will only drop pressure on liquids and must be liquid fed. All refrigeration expansion devices must be liquid fed as they will not drop pressure if fed by vapor refrigerant.

Ahead of all cap tubes you will find a small liquid line filter drier. The filter drier does two things:

■ It is supposed to keep the cap tube from getting plugged. The diameter of a cap tube may be no bigger than a pin hole.

■ The filter drier will absorb small amounts of moisture with its absorber to keep the cap tube from being plugged by ice crystals which will give intermittent cooling. The moisture will travel with the refrigerant and when it leaves the cap tube, the droplets of moisture become ice crystals and plug the cap tube. When the cap tube is plugged with ice crystals, the refrigeration process quits and the ice crystals melt. As the ice crystals melt, refrigeration starts again and the droplets of moisture travel through the system. When there are enough ice crystals to plug the cap tube again, this process repeats itself. To counter this problem, install a new liquid line filter drier and use a 24-hour vacuum or tripple evacuate the system.

The cap tube is able to drop refrigerant pressure by slowing down the flow of refrigerant. The diameter of a cap tube is very small. It goes from the size of a pin hole to just under a quarter of an inch. The length is 4 to 6 feet. The exception would be a thermostatic expansion valve that feeds a distributor which feeds several cap tubes. In this case, the cap tubes would be a shorter length.

After the refrigerant is slowed down, the cap tube enters the evaporator (which is about 10 plus times the diameter of the cap tube). The cap tube is expanding the gas volume by spraying the inside of the evaporator like a nozzle spray of a garden hose to a very large pipe. This refrigerant spray wets the evaporator all the way to the end.

The cap tube feed is constant and will not vary. There is no super heat with a cap tube. However, the evaporator that is fed from a single cap tube will be colder at the cap tube end and start getting a little warmer toward the suction end. This is why you will see a distributor with cap tubes that feed the top, middle and bottom of an evaporator.

What do you do about a plugged cap tube? First, make sure that you have a plugged cap tube. Attach the refrigeration gauges on the system and see if the suction goes into vacuum and the discharge becomes excessive and there is no cooling. When you have this condition, you know that there is an obstruction in the system. The obstruction could be in the driers, condenser or evaporator. Take a tubing cutter and cut the liquid line about 1 inch from where it is silver-soldered to the cap tube.

You will probably have no valves to blow the charge, so slowly cut the tubing. When you hear the hiss of gas escape around the cutter, stop cutting and let the charge escape. Another way to blow the charge is to take a hacksaw and gently cut the condenser pipe near the cap tube until you hear the hiss. Wait until the charge is gone. Use a tubing cutter to make the final cut. You will find that the cap tube might extend into the liquid line 2 to 4 inches because the manufacturer did not want silver solder to be drawn into the cap tube by capillary action.

At a convenient spot on the suction line, cut the suction line. Install a male, one-quarter flare on the suction that goes to the evaporator. Install a bottle of gas and charging hose on the flare and open the gas bottle valve. Take a match and light it and hold it near the open cap tube where the tube was cut at the liquid line. If the cap is plugged, the match flame will burn upright without a quiver. If the flame quivers or is blown out, the cap is not plugged. You have a bad liquid line filter drier unit, a plugged condenser, or an obstruction in the system.

WELDED SYSTEMS

When you are dealing with a system that is welded and you will have to use a hammer and chisel to open the back of the unit, you can use a cap tube cleaning tool. This tool is a small tube with a screw and a handle on one end and a nut and grommet that will fit the various sizes of cap tubes on the other end.

Unsilver-solder the cap tube from the liquid line or liquid filter drier. Be careful not to seal up the end with silver solder. If you have accidentally sealed up the end with silver solder, you will have to take a pair of gas pliers and cut the cap tube by fatiguing it. When the cap tube is small (pin-size diameter), use the pliers and bend the little tube back and forth until it breaks by metal fatigue. To get a clean break, use a knife or tubing cutter to score the outside of the cap tube. But do not go in too far. After the cap tube is scored, bend the tube at the score and it will break into a clean cut. Use the score and fatigue method on large caps. On small cap tubes, I use a knife to score the tube so the cut will be clean. Cap tubes cannot be cut by cutters or saws because the diameter of the tubes are too small and you would just seal the cap. When you cut it, score and fatigue the cap and try not to lose too much cap.

Place the cap tube end into the grommet and secure the cap tube on the cap tube cleaning tool.

Undo the screw handle on the other end of the tool.

Hold the cap tubule cleaning tool upright with the handle end at top and fill the tool with refrigeration oil.

Put the handle screw back on the tool and screw the handle screw all the way into the tool tube.

Repeat this process one more time.

Install a new suction line drier and make sure the refrigerant flow direction arrow is pointing toward the cap tube.

Clean the oil off the cap tube and insert the cap tube about 1½ inches into a new filter drier using a Schraeder valve. The Schraeder valve is needed to vacuum and change the system.

Use your pliers and bend the sides of the filter drier tube around the cap tube. Do not "kink" the cap tube.

Take a little silver solder and reconnect the cap tube to the drier and the evaporator.

Give the system a 24-hour vacuum or triple evacuate the system. See Chapter 11.

Charge the system using Chapter 9 pressures as a guide.

When you have access to both end of the cap tube, install a new cap tube and liquid line filter drier with a Schraeder valve. The Schraeder valve is needed to vacuum and charge the system.

TUBES WITHOUT ACCESS

I have had cap tubes I have not been able to unplug and with no access. In this case, I use the following precedures.

Cut a 1-square-foot hole in the back of the freezer or refrigerator near the top. Use a hammer and chisel and plus tin snips. Pull out the insulation, locate the cap tube where it attaches on to the evaporator, and remove the cap tube by unsoldering with torch or cut it out with a tubing cutter on the evaporator tubing.

Silver-solder the new cap tube on the evaporator. Run the cap down the back on the outside of the cabinet.

Hook up the cap tube to the new liquid line drier and the liquid line of the system.

Install the old insulation in the refrigeration unit.

Cut an oversized plate of sheet metal. By using sheet metal screws, attach the plate over the hole.

Use a little duct tape over the edges of the new plate to seal air leaks and you are in business. The back of the cabinets are against the wall so the plate and cap tube are not in sight.

What to do when you have a kinked or smashed cap tube:

Take a pair of gas pliers and remove the kink section by scoring, fatiguing and breaking the cap on each side of the kink. Remove as little as possible of the cap tube.

The system charge will slowly bleed out on either side of the broken cap tube.

Take a piece of copper tubing that is larger than the cap tube and cut it to a length of 4 inches.

Place the cap tube ends into the 4-inch copper tubing to where they extend in 1¾ inches each. The ends of the caps will be about one-half inch apart.

Using pliers, bend the copper tubing around the cap tube.

Take a soldering iron and lead solder and solder the connection. You may use silver solder if you wish. Do not sweat the solder too far into the sleeve connector.

You may use the process tube on the low side and install a Schraeder valve on the process tube. See I of Fig. 23-2.

Vacuum and charge the system.

If you want, you may place a pinch-off tool below the Schraeder valve and cut off the Schraeder valve with side cutters.

Using vice grips, flatten over the cap tube and fold it over.

Silver-solder the process tube.

The last step is to remove the pinch-off tool.

Cap tubes come in different sizes and lengths. They can be a little tricky to work with, but they are still one of the best expansion devices in the refrigeration field. Cap tubes can be purchased in 100-foot rolls. They also come in a kit all curled up with a new liquid line drier attached. When you buy the kit, the size of the kit is by horsepower or unit tonnage.

Most important is that all the cap tubes, because they are small in diameter, must be cut by scoring, fatiguing and breaking the tube.

Remember that when you use a knife or tubing cutter to score the cap tube, do not go in the tube too far and set up a burr or ridge on the inside of the tube.

In Fig. 23-1, letter "A" is the capillary tube. Letter B is the joint that is in the center of tube E. Both sides of tube E were flattened around cap tube A without altering cap tube A. The cap tube A extends 1¼ inches into the large tube E. This is the most common way to form a capillary tube joint either at the liquid feed pipe or at the evaporator pipe.

Letter E can be a liquid feed pipe (F) or cap tube (A) or the evaporator/cap tube joint. Letter C is a cap tube joint on the side of tube E. I just flattened tube E and contoured it around cap tube A.

Letter D is a sleeve coupling that joins a broken cap. Tube A1/A2 and cap tube A1/A2 were cut by score and fatigue and coupled by sleeve coupling D because of a kink.

Consider sleeve coupling D and cap Tube A1/A2 pictured to scale. A1 extends inside sleeve D to vertical line F and A2 extends inside sleeve D to vertical Line O. Between O and F is three-eighths to one-half inch. Cap tube A1 feeds liquid into sleeve D, which fills up and feeds liquid to cap tube A2, which feeds the evaporator with partical-size spray.

Do not use a lot of heat and solder at joint B or C. Just heat the joint and seal it with silver solder. You might get away with lead solder because there is no excessive heat or vibration at joint B and C. Extend all cap tubes into the joint of Pipe E or sleeve D about 1 to 1¼ inch before soldering. You do *not* want to seal the cap tube with solder.

In Fig. 23-2, letter A is a capillary tube. Letter E is a liquid feed pipe or evaporator pipe.

Letter D is a sleeve coupling for broken or kinked cap where the kink was scored and fatigued out. Letter I is a process tube that goes to the refrigeration system. A process tube is a one-fourth inch tube you will put on the refrigeration system so that the system can be vacuumed and charged. The one-fourth inch tube is pinched and closed with a pinch-off tool and the tube is cut between the pinch-off tool and the charging port on the end of the process tube. Then the tube is flattened, bent over and soldered. The pinch-off tool is removed in the last step. If you remove the pinch-off tool, before the process tube is soldered, you will always have a leak. A vice grip wrench would work for a substitute pinch-off tool.

The process tube is what the manufacturers of domestic units use in charging and vacuuming. It is the least expensive service entrance to a refrigeration system and it keeps you from repairing your domestic unit. There is no access valve because the manufacturers want you to buy a new machine.

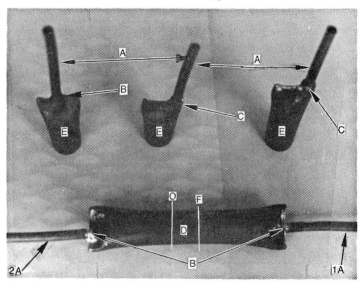

Fig. 23-1. A capillary tube and joints.

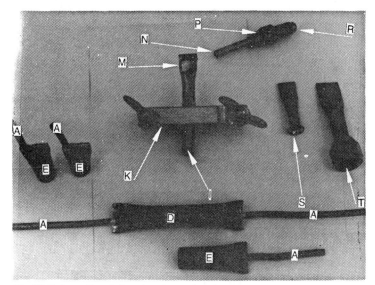

Fig. 23-2. A capillary tube, sleeves, and pipes.

Letter K is a pinch-off tool. It is a small wedge vice with wing nuts on either end. Letter M is a standard seal joining where the one-fourth inch tube was flattened and bent over. Letter N is part of process tube you save for the next vacuum and charge. Letter P is one-fourth inch flare nut. Letter R is one-fourth inch flare coupling that I use as service entrance port on the process tube. Letter S is a flared piece of one-fourth inch tubing. Letter T shows a flare nut over the flared tubing.

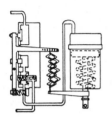

Chapter 24
Piping

With running copper tubing, whether it is hard wall straight or coiled, there are a few guidelines that will help. Lay out your components in the approximate location they are to be piped. Measure the lengths of tubes between components. When you need to change direction in close proximity, use a tubing ell or angle. Do not try to bend a close change of direction. The copper and the tubing will kink or the walls of the tubing will collapse. There are bending springs made for different size tubing. By keeping a constant tension on the outside diameter of the tubing, you might have luck in making a close bend without collapsing or kinking the tubing.

ANNEALING

If you are working with large diameter tubing and find it too hard to bend, take the piece of tubing you want to bend and heat it to a red metal colored heat. Then quench the hot tube into a pail of water. This is called annealing or metal softening. Do your bending immediately after the quenching. The quench operation can be repeated because copper hardens fast. If you still are not able to bend the large diameter tube, you will have to use ells, angles and offsets.

Put together the tubing and components, but do *not* solder. See how everything fits. Make adjustments, if you are short, by using some of the space inside of the ells or couplings.

When you are using a substitute compressor, make sure you have correct mounting dimensions for the new compressor feet. Otherwise, purchase an adopter plate so the new compressor can be mounted. On a new compressor, try to make sure that the discharge, suction, oil coller and process tube pipes have approximate placement to the old compressor pipes. Otherwise, you are going to have a transition work or rerouting of tubing.

Clean up the components where you will be silver soldering. Use fine sandpaper, emery cloth, or a fine file. If the components are copper to copper, use Still Flos silver solder. If the components are copper to brass or steel, use Easy Flow 45 + Borax flux.

Make sure that all pipes and components that are hard to get at are soldered first. The technique to use is to apply medium flame to a joint. The flux will turn white and then to a clear water color. The copper tubing will turn from copper color to black to red. When you have the red color, melt one little bead or silver solder bead ball on the joint. Keep applying the heat or the flame up and down the joint until the ball melts. When this happens, apply the silver solder and it should flow easily into the joint. Apply the flame on the top and bottom of the joint at an angle so that the flame will lick around the joint. Touch a little silver solder around the top and it will flow on the underside without your ever touching the torch to the back side of the joint. You will have to use this technique many times because the working quarters are too small and there is no room for the torch inside.

You will have to solder the suction line drier and piping that is behind the compressor outside the system. Place the compressor on a table and prefabricate and silver-solder all inaccessible joints and difficult joints. Then install the compressor with soldered components. Have the last two joints accessible because it is nice to solder with minimum difficulty. Make sure that the silver solder you use is cadmium free. Cadmium is a very toxic poison and will give you trouble if inhaled or breathed.

Use a wet rag around and over parts that are sensitive to heat. Examples are thermostatic expansion valves, capillary tubes, solenoid valves, driers and sight glasses. Too much heat will break the glass crystal of a sight glass or burn the molecular sieve inside a drier. Too much heat could warp the moving parts of valves. Because the cap tube diameter is so small, you could melt it. Use care in silver soldering. When the wet rag becomes dry and you are not finished, take the rag off, rewet the rag, and wrap it around the work again before you continue silver soldering.

Generally, copper tubing is good for 2000 pounds per square inch of pressure so it will handle all piping requirements of refrigeration, air conditioning and heating. Copper tubing is not used on ammonia systems because ammonia will eat and erode copper. Copper tubing is not used on natural gas lines. A chemical reaction develops and the inside of the copper tubing will fill up with a black flaky substance and the entire line will be plugged up in short order. Look at the pilot and connecting lines for gas fired furnaces. Most of them are aluminum tubing.

COMPONENTS

In Fig. 24-1, letter A is a flare vice. Letter B is a flare union coupling. Letter O is the bevel of a standard flare. Letter P is a flare nut.

In Fig. 24-2, letter A is a standard compression connector that can be used quite successfully in joining tubing of the same diameter. Letter B is a

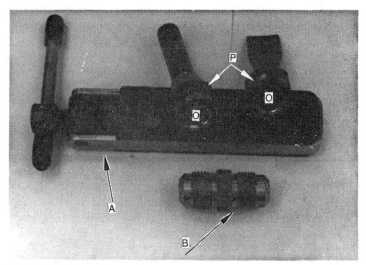

Fig. 24-1. Flares and a coupling.

ferrule that is needed to form a seal between the tube C1 and the connector body. Letter C is the connector nut that secures the tube, ferrule, and connector body. Letter H is a nut that goes on a special compression connector called a swageloc. This compression connector has two ferrules. Letter F is the main ferrule. Letter G is the keeper furrule. When you have a leak in a pipe or need to join two pipes of the same diameter, this can be done using compression connectors. Consider using compression connectors, because a welding torch could give you problems. You might not have enough working space and a torch is to cumbersome.

Letter A of Fig. 24-3 shows the body of a standard compression connector. Letter B is the ferrules. Letter C is the compression connector nuts. Letter D is a flare union coupling. Letter E is the flare nuts that will hold the bevel of the tube on to the seat of the flare union coupling. Letter F is the main ferrule. Letter G is the keeper ferrule.

Any coupling that has the two ferrules, F and G, is called a swageloc compression connector. Letter H are the coupling nuts that hold tubing, compression body and ferrules together.

My first choice is the standard compression connector A, B, C. This is simple to make up and all you need are two crescent wrenches: One for the nut (C) and one for the body (A). The swageloc compression connector (F, G, H) is fair, but there are two ferrules instead of one. The flare compression connector is the most commonly used. However, the flare compression connector requires a flare vice and flare tools as well as two crescent wrenches.

In Fig. 24-4, letter C is a standard compression connector. Letter E is a flare union coupling. Letter H is the swageloc coupling. This swageloc coupling H is a 90- or 80-degree ell. You can get any of these couplings in

Fig. 24-2. A standard compression connector.

tees or 90-degree ells and sometimes in 45-degree ells. Using these connectors, it is possible to put a whole refrigeration system together without a welding torch. The only problem is the cost of couplings.

In Fig. 24-5, letter A is the flare vice. Letter O is a piece of copper tubing. Letter B is a swedging tool. A swedging tool is a tool that will make

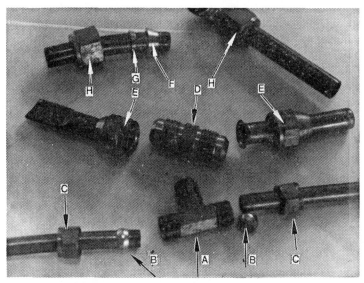

Fig. 24-3. A standard compression connector.

Fig. 24-4. A standard compression connector (C), a flare union coupling (E) and a swageloc coupling (H).

the inside diameter of tubing the same size as the outside diameter. For example, suppose that tube "O" and "P" are the same diameter tubing. You can ram swedge B into the inside of tube O. Next, remove swedge B from tube O. Then place tube P inside of tube O where the swedge was and solder this connection with a torch. This is quite standard in use. When this coupling is soldered, it is called a *swedge coupling*.

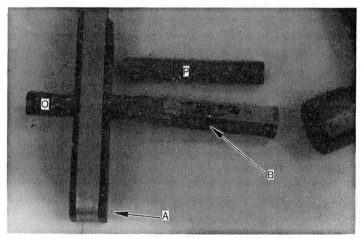

Fig. 24-5. A flare vice and copper tubing.

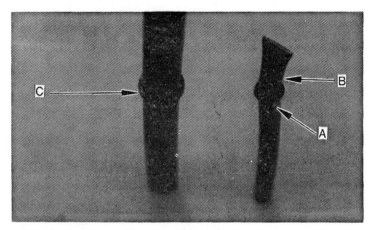

Fig. 24-6. A flare vice and copper tubing.

Figure 24-6 shows a tubing coupling I have used with success. This is seldom used, but it will work when you do not have materials and tooling to couple the tubing any other way. Letter A is a flared piece of copper tubing. Letter B is a straight cut-off piece of copper tubing. Butt up the straight piece (B) into the flared piece (A). Heat both pieces of copper (A and B) with a torch and fill the exposed flare area with silver solder. Letter C is a quite strong joint and can take vibration when the little well of flare A between B is filled with silver solder. I like to call this tubing joint a butt flare coupling.

Figure 24-7 shows soldered tubing joints. Take the straight cut tube (B) and place it in the flared tube (A). Next, heat A and B with a welding

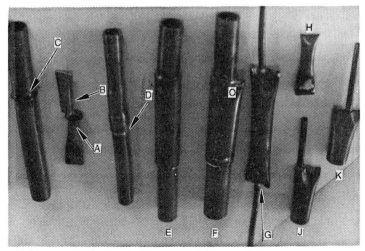

Fig. 24-7. Couplings.

torch and apply silver solder in the little flare well at letter C. Try to fill the well with solder by applying some heat to the silver solder rod as well as the tubing.

Letter C is a butt flare coupling. Letter D is a standard swedge coupling. Letter E is a sleeve coupling. Take a tube that will just fit over the two tubes (E) and silver-solder both ends of the sleeve.

Letter F is a sleeve coupling that is joining two tubes of different diameter. Fit the sleeve so that it is snug fit to the largest tube that you want to join. Place a vice grip wrench or gas pliers at letter O and crimp the sleeve around the smaller tube. Silver-solder both ends of the sleeve.

Letter G is a small sleeve to repair broken or kinked cap tubes.(See Chapter 23 for how to cut a cap tube. Insert the ends of the cap tubes into the sleeve about 1¼ inches).Take a pair of gas pliers and crimp the sleeve around the cap tube. Be careful not to collapse the walls of the cap tube. Silver-solder both ends of the sleeve. Letters J and K are cap tube joints. The cap is either joined to the liquid line or the evaporator. Insert the cap tube into the liquid line or the evaporator 1¼ inches. Take a pair of pliers and crimp the large pipe around the cap. Be careful not to kink or collapse the cap tube. Silver-solder the cap tube joint.

Letter H is a tubing seal. Take a pair of pliers and flatten the tubing for one-half inch. Roll over and flatten the flat tubing on itself. Apply solder on the tubing end.

With Fig. 24-8, imagine that tube F and A are one continuous tube and that tube AF is a process tube that tees into the suction. Letter X goes to the low side (suction) of a refrigeration system. Letter C is a one-fourth inch flare coupling held onto the process tube AF by flare nut B. Letter C is the service entrance port to the vacuum and charge a refrigeration system on the low side. When you are finished with the vacuuming and charging of the system, place a small pinch vice, called a "pinch-off," at G. The jaws of a vice grip could work as a pinch-off at point G. You must pinch the tubing at point G to valve off port C so you can remove gauges without blowing the charge.

At letter H cut the process tube. Next, flatten the tube at point H like you see at letter D. Letter E shows rolled over flattened tube. Next, letter H is heated with a torch and soldered. You can use either silver solder or lead solder on this tube seal. It is very important not to remove the pinch-off at point G (which is between the refrigeration system X and the service port C) until letter H has been soldered. If the pinch-off is removed before letter H has been soldered, there will be a minute flow of refrigerant through point G and there is no way to seal a pipe with solder under gas pressure. Even a drop of gas refrigerant will cause a pin hole in the solder that you might not see, but there will be a leak and all your work will be for nothing.

You can remove the gauges from a process tube after you have vacuumed and charged the system without blowing the charge. Use the following procedure:

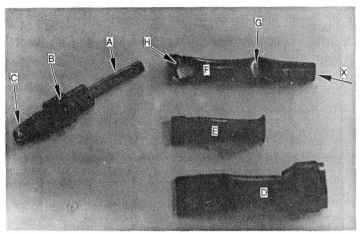

Fig. 24-8. Tubes.

■ Apply the vice grip or pinch-off vice at point G. By doing this, you have valved off port C where the gauges are attached.

■ Remove gauges from port C.

■ Take a flare cap that has a copper or rubber washer inside of it and install the flare cap with washer on port C.

■ Remove the vice grip or pinch-off vice from point G and you are finished.

The process tube is the principle way the manufacturers gain service entrance on domestic air conditioning and refrigeration equipment. You can do the same if you use my technique.

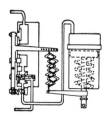

Chapter 25
Charging a Unit

At least twice a week someone will ask me how to charge their refrigerator, freezer, or window-rattling air conditioning unit. When you closely look at a domestic unit, the first thing you will discover is that there are no service entrance valves. The manufacturers have done the vacuum and charging using process tubes. If you try to use the same tubes, you will get involved with vacuuming, charging, and tubing work. All you want to do is put a little refrigerant gas in the domestic unit with no other problems.

MATERIALS LIST

The following is a list of low-cost charging materials.

One adjustable line tap valve with a one-fourth inch flare-cap.

One small can of refrigerant.

One Tap-A-Can® valve.

One small compound gauge: 0-30 inches vacuum, 0-150 psi, 0-200 psi, one-eighth inch pipe thread or 0-250 psi; J-B, M 1-250.

One female coupling.

One flared brass tee (all flares ¼-inch).

Six one-fourth inch flared coupling nuts.

Three small lengths of one-fourth inch copper tubing (two 6 inches long and one 4 inches long).

On top of a R-12 refrigerant can which has a gas weight of 15 ounces, install a tap valve. On most small domestic units, you are dealing with less than two pounds of gas. You may buy a 16-ounce can of R-22 gas if the domestic unit is R-22. The type of refrigerant that any refrigeration system uses, is always printed on the name plate tag. Generally, the manufacturer will tell you the volume in ounces of refrigerant that the machine will hold as well as the name of the refrigerant.

Letters E, F, G, and H of Fig. 25-1 are the line tap valve that has been broken down into its components. This particular line tap valve can be used with three-eighths, one-half and five-eighths inch copper tubing. All tubing is measured by the outside diameter.

Letter E shows the inserts that will allow you two additional sizes of tubing. If you have five-eighths OD tubing, just use the saddle clamp (letter F), and mount the saddle clamp and valve, letter H, around the copper tube. For smaller tubing, such as three-eighths or one-half inch, place the appropriate insert inside the saddle clamp and then amount the saddle clamp (F) and valve (H) on the low side suction of the refrigeration system.

Letter H is the body of the piercing line tap valve. All line-tap valves are known as saddle valves. They mount around the pipe or tube like a saddle and cinch on a horse. Hence, the name saddle valve applies.

Letter G is a one-fourth inch flare cover cap. Cover the service entrance valve port of valve H with the flare cover cap G when you have completed charging your refrigeration system.

In Fig. 25-1, letter A, I have made three little copper hoses using the flare nuts and copper tubing. Letter B is the flared brass tee. Letter C is the female coupling, male flare one-fourth inch, female pipe thread one-eighth inch. Letter D is the compound gauge. This one goes from 0-30 inches mercury vacuum and 0-250 psi.

Figure 25-2 shows a line tap valve that I have taken apart to show its components. Letter E is the inserts for different size tubing. Letter F is the valve saddle clamp. Letter H is the valve body. Letter J is the valve needle that will pierce the suction tube when you turn the valve handle L

Fig. 25-1. Refrigeration charging components.

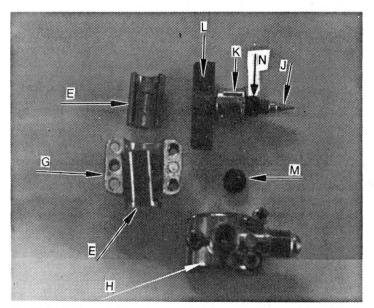

Fig. 25-2. Line tap valve components.

all the way inward. Letter K is a nut that will tighten packing N into the crown of valve body "H" and secure the valve needle assembly J. Letter M is a little rubber gasket that fits in a well on the underside of valve body H.

Before you install the line tap valve, make sure the little rubber washer is in the well. When you turn the handle so that the needle goes into the tubing, the needle will go through the small hole of the valve washer and into the tubing. The needle makes a tapered hole in the tubing and this tapered hole is the valve seat. When you open the valve, the needle moves away from the seat and the refrigerant can flow from the refrigeration system through the little tubing valve seat hole, through the gasket (Letter M), and through the service entrance port of valve body (H).

Make sure nut K is tight and valve saddle clamp F is secure on the tubing. Otherwise, you will have a leak at the packing (N) or washer (M). If the washer falls out and you install the line tap valve and turn the handle in—so the needle will pierce the suction pipe—you will not be able to use the valve. All the refrigerant will leak out of the washer well. It is imperative that washer M be in its little well. As shown in Fig. 25-3, I have cleaned off the low-side suction tube with fine sandpaper so it is clean where the valve mounts. The clean copper tubing will help the valve washer M make an excellent seal and valve needle J to make an excellent seat in the copper tubing.

Letter F is the valve saddle clamp which is under the low side suction tube (S). Letter H is the line tape valve body. If you look carefully, you can see the washer (M) in the little well in the inside curve of the body of the

210

valve that will be clamped on to the tubing. There are three little Phillip screws that are to the right of valve body H and these screws will secure saddle clamp F to valve body H.

As shown in Fig. 25-4, I have cleaned the suction line (S) with fine sandpaper and mounted the line tap valve (H). The valve handle has been turned clockwise and the tubing has been pierced. On the service entrance valve port, I have secured the little copper tube hose (A). On the end of the copper tube hose, secure the flare union coupling and on the coupling secure hose connector N and place the gauge D on the suction line. The hose connector (N) has a little indenter inside that can be used with Schraeder valves and will depress the valve cores to gain service entrance.

In Fig. 25-5, the line tap valve (H) is on the low side suction pipe (S). When you mount the line tap valve, make sure the valve handle can be turned without problems and that you have enough clearance around the service entrance port to secure the little copper tube hose (A).

Letter C is a female flare coupling, (one-fourth inch male flare, one-eighth inch female pipe thread). Letter N is a charging hose and pipe connector (one-fourth inch female flare, one-eighth inch female pipe thread). Letter O is a one-eighth inch male pipe nipple thread and I have put pipe dope on it so that when I secure the nipple O into either connector N or C, there will be no leak. Letter D is the compound gauge.

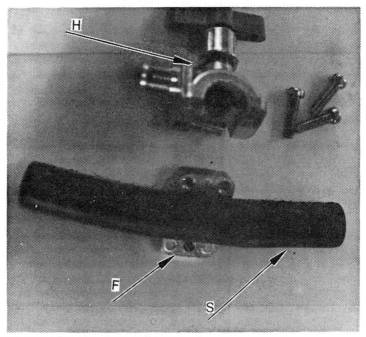

Fig. 25-3. The low-side suction tube.

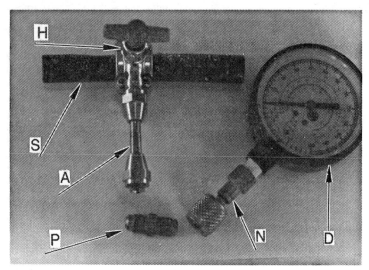

Fig. 25-4. A suction tube and compound gauge.

In Fig. 25-6, I am showing you the hardware you will need to put a gauge on a refrigeration system. Letter S is the suction line. Letter H is the line tap valve. Note the saddle clamp F on the underside of the suction line.

1. Make sure the line tap valve handle is all the way in.

2. Remove the flare cap (G) from the line tap valve service entrance port.

3. Secure the little copper tube hose (A) to the line tap valve service entrance port.

4. Secure the female flare pipe coupling (C) to the little copper tube hose flare and gauge pipe nipple O.

5. Open the line tap valve and read the compound gauge D.

An alternate method is to first do it this way: do number 1 and 2 above and then:

3. Secure the charging hose pipe connector (N) to the compound gauge (D).

4. Secure the charging hose connector (N) to the service entrance port of line tap valve (H).

5. Open the line tap valve (H) and read the compound gauge (D). Once you have secured C or N to the gauge, you may leave the connector permanently attached.

In Fig. 25-7, I am starting to build up the charging gauge manifold system. The line tap valve (H) is mounted on section line (S). Copper tube A is secured to the service entrance port of the line tap valve (H) and the flare T (letter B). The gauge (D) is connected to the flare T (letter B) and the other little copper tube (A) will connect onto the flare T (letter B).

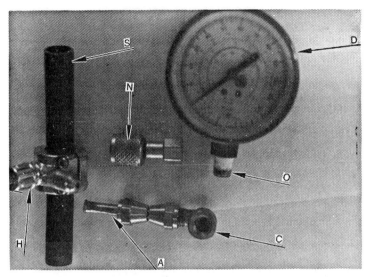

Fig. 25-5. The line tap valve is on the low-side suction pipe.

In Fig. 25-8, you can see the assembled charging gauge manifold (letters D, N, X, and H1). Instead of using fitting N for connecting gauge D to the flare T, you could use C and the little copper tube hose.

The line tap valve (H) is secured on suction line (S). Flare cap G is secured on the service entrance port of line tap valve (H). During the charging operation, letter X will secure on the service entrance port of line tap valve (H) and letter H1 will secure on the tap can refrigerant valve port.

Fig. 25-6. Hardware needed to install a gauge on a refrigeration system.

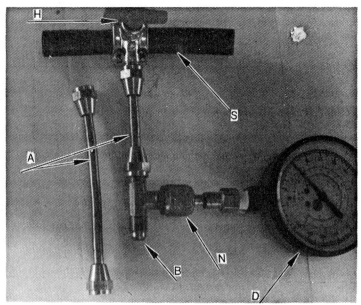

Fig. 25-7. Building a charging gauge manifold system.

Figure 25-9 shows the completed charging gauge manifold. Letter A are the little copper tube hoses with flare nuts. Letter B is the one-fourth inch flare T. Letter C is the pipe flare coupling (one-fourth inch male flare, one-eighth inch female pipe thread. Letter D is compound gauge (D).

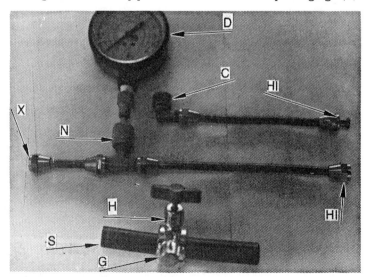

Fig. 25-8. An assembled charging gauge manifold system.

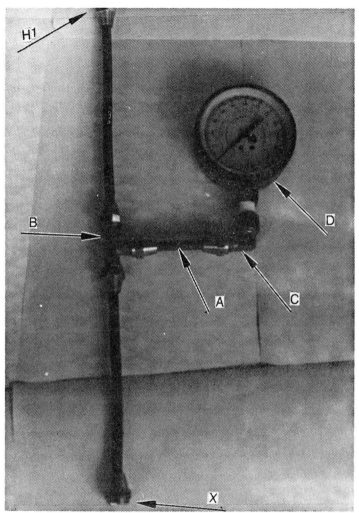

Fig. 25-9. The completed system.

Letter X will secure to the service entrance port of the line tap valve and letter H1 will secure to the tap can refrigerant valve.

With my system, there is a service entrance valve and port, a charging manifold and gauge, a refrigerant valve, and Freon refrigerant in a 15-ounce or 16-ounce can. You can charge any R-12 or R-22 refrigeration system.

Figure 25-11 shows the complete Roger Fischer charging gauge manifold system with a line tap valve and its service entrance port at letter X and the tap can valve with R-12 refrigerant which will connect by flare nut at letter H1.

The tap can valve (letter H1) is a piercing valve similar to the line tap valve. This tap can valve has a female one-eighth inch pipe thread that screws on to a male one-eighth inch pipe thread that is on a stud of the R-12 refrigerant can. To attach a valve to the can, first make sure the little gasket is in its well on the underside of the valve and that the handle is unscrewed all the way out. Screw the tap can valve on the male stud of the R-12 refrigerant can. Turn the valve handle all the way until it will turn in no farther. At this point, you have pierced the top of the male stud. When you open the valve, refrigerant will flow out of the valve port (letter H1).

Letter S is the suction line. Letter B is the one-fourth inch flare tee. Letter N is the pipe-to-charging-hose connector with an indenter for Schraeder valve cores (one-fourth inch female flare and one-eighth inch female pipe thread. Letter D is the compound gauge (0-30 inches mercury vacuum and 0-250 psi pressure).

CHARGING PROCEDURE

The first thing you must do is to locate the name tag of the unit you are going to charge and find out the type of refrigerant the unit has. Is the unit R-12 and R-22 refrigerant? Most small domestic refrigeration units will have less than 2 pounds of refrigerant inside the system. Next, take a materials list with you and pick up the parts from a local refrigeration supply house or retail store. You might get an extra can of refrigerant just to be on the safe side.

Remove the back or side cover of the unit and expose the compressor. You must determine which of the pipes that leave the compressor is the low side or suction. If the compressor is a derby-like can, the top half of the can will have a process tube (this is a one-fourth inch tube, about 4 inches long, that extends from the can and dead-ends or is sealed off) and the suction pipe or low side.

You do not want the pipes that leave from the bottom of the Derby can. These pipes will either be discharge pipes or oil cooler pipes. If the compressor is horizontal, the largest size pipe will be a suction pipe if it goes inside of the refrigerator or freezer compartment. When the pipes are the same size, feel the pipes. One pipe will feel hotter than the other when the compressor is running. This hot pipe is a discharge pipe and you are feeling the heat of compression. When you feel the pipe, feel it about 2 inches from the compressor can. The other pipe is the suction pipe or low side pipe.

There is another way to find the suction pipe or system low side. Observe both pipes that leave the compressor. Follow them as far back as you can. If the little capillary tube is soldered on the outside or is run inside one of the compressor pipes, this is the low side or suction pipe. Many manufacturers will run the capillary tube parallel to the suction and lead solder it to the outside of the suction pipe as a heat exchanger. You have a tube on a tube heat exchanger and are precooling the refrigerant feed to the evaporator by the cool suction return line.

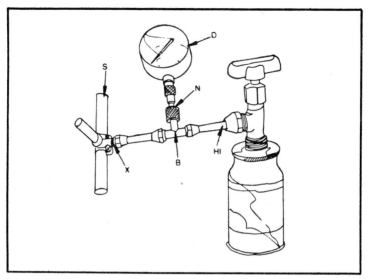

Fig. 25-10. The gauge manifold system.

If the manufacturer runs the capillary tube inside the suction, he is forming another heat exchanger—which is a tube in a tube—and the result is a pre-cooling of the refrigerant feed to the evaporator by the cool suction return line. What the manufacturers have done by using these heat exchangers looks good in theory, but in practice the suction line is not cool and I do not think any pre-cooling takes place.

If you need to get on the suction line and you have the cap tube soldered on the outside of the suction, you can take a small butane or prepto torch and unsolder the cap tube very easily. The manufacturer has used lead solder and the cap can be unattached and the pipe cleaned with a little heat and rag. Do not touch the tube in a tube heat exchanger. Place your piercing line tap valve in an accessible place on the suction between the heat exchanger and the compressor. It should be at least 2 inches away from the heat exchanger, if possible.

When you have an outside static coil or coils, the job is now very easy. *All* pipes that go to outside coils are not used. The pipe that leaves the compressor and goes to the inside compartment of the refrigerator or freezer is the suction and this is the pipe you are going to work with. After you have identified the suction pipe or the low side pipe, you will install the line tap valve in place on the pipe where the pipe is most accessible.

On all of the illustrations in this chapter, there is a letter S on a piece of copper tubing and this is the suction line of the machine. I installed the line tap valve on this suction line S. There is another way to install the line tap valve and keep the valve and use it over and over again.

Start at where the suction leaves the compressor and follow the pipe all the way back to the evaporator (cooling coil). Some place there is a one-fourth inch tube about 4 to 6 inches long that is tied into the suction line and this tube, which is 4 to 6 inches long is sealed off and goes nowhere. This is the process tube.

The manufacturer used this one-fourth inch dead-ended tube to vacuum and final charge your machine. If you can not find this tube on the suction line, check the compressor shell and there is normally a process tube on the shell or can of your compressor. All process tubes that are on cans of full hermetic compressor are suction or low side.

Get a pair of vice grips. Purchase a line tap valve that will work for one-fourth inch tubing. Mount the line tap valve nearest the sealed dead-end of the process tube. You must have enough length on the process tube between the line tap valve and the suction line of the machine so that you can pinch the one-fourth inch process tube closed with a vice grip or pinch-off vice. The procedure goes like this:

■ Mount the line tap valve on the process tube.

■ Charge the machine and keep the hose attached to the line tap valve.

■ Pinch off the process tube between the suction line and line tap valve. The pinched process tube valves off the line tap valve.

■ Remove the line tap valve.

■ Take a propane or prepto torch with lead solder and seal the little hole that was made by the line tap valve.

■ The last thing you do is remove the pinch-off. Do not remove the pinch-off until you have soldered the little hole made by the line tap valve. Otherwise you will have a leak. You are doing the same thing with the line tap valve as the manufacturer does with a process tube male one-fourth inch flare. In doing this little trick, you can use the line tap valve over and over. Do not do this procedure if the process tube is not accessible or too short. You will have to use your line tap valve on the machine suction line.

To install the line tap valve, first take sandpaper and clean the surface of the suction pipe where you are going to install the valve. Remove paint, dirt and any roughness or burrs on the suction pipe surface. Check and make sure the little rubber gasket is in place where the piercing needle comes out of the valve on the underside. Mount the saddle-type valve on the cleaned suction pipe so that the valve charging port is accessible and you can turn the handle without getting hung up. Make sure the screws that hold the saddle-type valve are tight and the little rubber gasket is seated on the suction pipe.

Next, turn the valve handle until you can feel it a little hard. At this point give one-half a turn and open the valve and listen for gas hissing out of the valve port. All you do is make a little pin hole, with a piercing needle, on the line tap valve. Do not collapse the copper suction pipe. Use care and only go in with the handle of the line tap valve a half or one-third of a turn at a time and back off and listen for hissing of gas out of the port.

When you have a hiss, turn the valve handle in until the hiss stops and there are no more turns of the handle. Once the line tap valve is installed, it cannot be removed without blowing the charge. Then you would have to repair the hole in suction line, install a service entrance valve or a process tube, vacuum and charge system.

The reason I went to such detail to help you find the suction pipe is that if you put a line tap valve on the oil cooler pipes or a discharge pipe, you will have to turn off the valve, put the flare cap over the charging port, leave the valve on the line, and buy and install another valve on the suction pipe. You cannot charge on the oil cooler or discharge pipe.

Assemble the parts that you have purchased so that your layout is the same as in Fig. 25-10.

■ The line tap valve (H) is turned off or closed.

■ Set the refrigerant can upright and open the tap can valve (letter K) and fill the system with vapor refrigerant.

■ Unscrew the flare nut a little bit (at letter X) and let the vapor refrigerant escape from the loose flare for two seconds. Then secure the flare (letter X) to stop refrigerant leak. In doing this step, you have purged the system of air that was inside and filled the system with vapor refrigerant. Every time you charge a system, you must get rid of the air and moisture in the gauge and manifold by the purge method in this step #3.

■ Start the domestic unit and close the tap can valve.

Before you add refrigerant to the refrigeration system, see Table 25-1 for the pressures that the system will charge up to.

With the engine running at fast idle and with average heat load, look at the sight glass and charge on suction until bubbles in the sight glass go away and sight glass stays clear. The sight glass is between the condenser and the thermostatic expansion valve. The sight glass can be in line or integrated with the drier. When there is no sight glass, consider 20 psi with the motor running quite fast and up to 40 psi with motor running at slow idle. Table 25-2 lists compressor speed changes system tonnage and pressures. With the compound gauge, start charging the refrigeration system.

■ Open the line-tap valve and read the gauge after it settles down. If after 10 minutes of running, the gauge on the suction line reads 100+ psi for a R-12 system, or 200+ for a R-22 system, shut the system down. You have bad compressor valves and a system charge will accomplish nothing. To repair, you have to change the compressor, vacuum, and charge system.

■ With the refrigeration system running, read the suction pressure. Open the tap can valve and you can hear the refrigerant leave the little can and go into the refrigeration system. Have the refrigerant can standing upright. When the can is standing upright, you are putting the refrigerant into the system as a gas or vapor. If you turn the refrigerant can upside down, you are putting refrigerant in the system as a liquid. All refrigeration compressors are vapor pumps. Liquid refrigerant will break the fluttering reed valves of the compressor. Try to do most of the charging with vapor (refrigerant can standing upright).

■ After one-half minute, turn off the tap can valve and watch the gauge. In a minute or two, the gauge pressure will drop and stabilize (compare the machine pressure to chart pressure).

■ Repeat this process of turning on the tap can valve for one-half minute or more and turn the valve off and let the gauge stabilize (compare the machine pressure to chart pressure).

■ If you want to speed up this charging process, you can momentarily turn the small refrigerant can upside down a few times and turn off the tap can valve and wait for the gauges to stabilize.

■ When the machine is starting to cool down and has been running for an hour, your gauge pressures should correspond to the chart pressures (Table 25-1). In case of a freezer or a refrigerator, I would leave the refrigeration setup on for four to six hours and check to see that you maintain the stabilized chart pressures.

■ Turn off the line tap valve and tap can valve and undo the flare nut (at letter X in Fig. 25-10). Remove the charging system and place a flare cap (G) on the service entrance port a letter (X). Undo the flare nut at letter H1 and you can store the charging system.

■ If you have placed in two cans of refrigerant and the gauge still reads in inches of mercury vacuum, shut the system down and remove charging system. There is an obstruction in the refrigeration system. The blockage is in the drier, cap tube, strainer of expansion valve, condenser or evaporator. You will have to find and remove the obstruction, vacuum the system and final charge according to my pressure chart (Table 25-2).

If you have reached the correct pressures and then go into vacuum, from time-to-time moisture in the refrigeration system causes an obstruction when it freezes and blocks the system with ice on leaving the cap tube. The ice crystals that plug the cap tube send the gauge into 20-inch mercury vacuum and the refrigeration unit becomes warm. At this point, the ice crystals melt, unite with the refrigerant, flow through the system, and eventually plug us up again. Install a new liquid line filter drier, vacuum, and final-charge the system.

After you charge the refrigeration unit, take a vacuum cleaner and remove all the dirt, dust and lint from the oil cooler, condenser, condenser fan, and compressor can. I have removed many a dead mouse from the condenser section of a freezer or refrigerator. The mouse will crawl in

Table 25-1. Refrigeration System Pressures.

On R-12 freezers, final suction pressure, cold box	=	1-2 psi
On R-22 freezers, final suction pressure, cold box	=	10-13 psi
On R-12 dual temp. refrigerators, final suction pressure, cold box	=	6-8 psi
On R-22, dual temperature refrigerators, final suction pressure, cold box	=	20-22 psi
On R-12, Window ac Units, final suction pressure	=	37 psi
On R-22, Window ac Units, final suction pressure	=	69 psi
On R-12, Auto/Truck Air Conditioner, final suction pressure	=	20-40 psi

Table 25-2. Variable Compressor Speed Changes System Tonnage and Pressures.

On R-12 heat pumps, charge on cooling, suction pressure	= 37 psi
On R-22 heat pumps, charge on cooling, suction pressure	= 69 psi
On R-12 dehumidifiers, suction pressure	
If evaporator coil has ice, raise pressure to 26 psi	= 22 psi
On R-22 dehumidifiers, suction pressure	= 44 psi
If evaporator coil has ice, raise pressure to 50 psi.	
On R-12 coolers for milk, beer, and fluids, suction pressure	= 37 psi
On R-22 coolers for milk, beer, and fluids, suction pressure	= 69 psi

where the warm condenser is to get away from the cold outside. The condenser fan starts and it is good-by mouse. By having the condenser section clean, the efficiency of your machine could be improved 10 to 15 percent.

Dirt acts as insulation and it will impair and lower the efficiency of the condenser and oil cooler. The unit gets rid of the heat that is inside a refrigerator, freezer, and beer or milk cooler by the condenser. The heat leaves the refrigerant and travels through the metal of the condenser and unites with the air around the condenser. The three methods of heat transfer take place at the condenser: connection, conduction, and radiation. Many times, a small fan will aid in the heat removal. When the condenser is insulated from the outside air by dust, dirt, and lint, the head pressure or discharge pressure of your system will be much higher. Many times you will find insulation on the walls inside the compressor compartment. The manufacturers use this insulation to soften the compressor noise. When this insulation becomes unglued, falls down and obstructs air flow, remove this insulation and put it in the trash. I doubt if you will notice any increased noise level.

The second thing you can do is check the door gaskets of the machine you have charged. When the door gaskets leak, the machine might always be running, but it will not get cold enough. There are two ways to check a gasket. Put a lit flashlight inside your machine and turn off the room lights. Check to see if there is any light between the door gasket and door jam. The second way to check a leaky door gasket is to open the door of your unit and place a dollar bill between the door jam. Close the door. The dollar bill should stay in place and very hard to pull out. When the unit fails, the flashlight or dollar bill test, try this next. Remove covers that are over the door hinges. Move the hinges in one-fourth inch. Move the door catch in one-fourth inch. Doing this will cause the gasket to stop leaking cold air or slow the leak down. You also can take newspaper and roll it up and place it inside the gasket. By doing this, you are expanding the gasket to attempt a snug fit of the door gasket to the jam.

When you take the time to charge a refrigeration unit, spend just a little more time and check to see that the compressor condenser unit is clean, the door gaskets do not leak, and the door hinges and latch are in good repair.

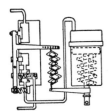

Index

Edited by Steven Bolt